THE KARRI FOREST

Its conservation significance and management

DEPARTMENT OF CONSERVATION AND LAND MANAGEMENT

THE KARRI FOREST

Its conservation significance and management

Compiled and written by P.E.S. Christensen

Edited by Marianne R.L. Lewis

A CALM Publication

Published by the
Department of Conservation and Land Management
Como, Western Australia

ISBN 0 7309 4676 2

February 1992

Cover Design: Stacey Strickland
Photographs: Jiri Lochman
Bill Bachman

ACKNOWLEDGEMENTS

This publication has been some time in the making and many people have contributed. First of all, I thank Andrew Burbidge and Roger Underwood for originally suggesting the publication, and for their suggestions on the structure and content. Roger also contributed to the Introduction and commented on several drafts. I am also grateful to Ian Abbott, Charles Dortch, George Gardner, Paul Jones, Barbara and Bert Main, Neville Marchant, Norm McKenzie and Alan Walker for critical comments on the text and in some cases also editorial suggestions.

Tony Annels deserves special mention for the many hours he put in assisting with compiling information on plant species and vegetation associations of the area. Many others too numerous to list have assisted in providing a sounding board for ideas and helping with queries on technical matters.

I thank Joanne Healey and Kerry Carmichael for patiently word processing several versions of the text and Marg Wilke for the desktop publishing.

Barney White, a private consultant and former forester in the southern forests was contracted to provide the data for the historical section and the initial draft of the section on forest management.

I am also grateful to staff of Land Information Branch CALM for preparing the Figures.

NOTES FOR THE READER

In providing the detail which is often necessary in this text, many place names not ordinarily shown on maps of the area, have been used. It has not been possible to include a map with this level of detail so the interested reader is referred to CALM's 1 : 63 360 scale maps (available from the Department of Conservation and Land Management, P.O. Box 104 Como, Western Australia).

Throughout the text, especially where reference is made to biogeographical information, the term 'south-west' has been used.

In Western Australia there are three distinct climatic regions. These are the tropical north with a short hot rainy period, and the arid interior which has no regular rainfall pattern and the temperate south or south-west with a rainy period occurring in the coolest months. The flora is remarkably different within these areas which, for this reason, are often used as broad natural divisions of the flora.

There exist many definitions of the south-west, the one which most approximates the area which is referred to in this text is the one described in the West Australian Yearbook (see Sayer 1969) viz.

> The South-West Province extends from the southern end of Shark Bay in the north to Israelite Bay in the south. On the west and southern sides it is bounded by the ocean, while its inland boundary passes close to Mullewa, Morawa, Koorda, Bencubbin, Burracoppin, Hyden, Ravensthorpe and Grass Patch. It is pre-eminently the winter rainfall province which receives its maximum rainfall from May to August inclusive and, with the exception of the southern portion, experiences a seasonal drought extending from November to March or April. The average maximum temperature is less than 80° F (27°C) with much lower temperatures during the growing season.

This distinction of a south-west province or region is also useful when dicussing fauna distribution. It is approximately equal to the Bassian region which is sometimes used by bird biogeographers (see Serventy and Whittel 1976). The distinction of a south-west province or region also has meaning for some of the lower phyla, the more mesic element like the amphibia, freshwater fish and the aquatic invertebrates.

Lastly, there have been difficulties in referring to some national parks and nature reserves because their status is in transition from one tenure to another. Land tenure and classification arises from a process of Regional land use planning subject to eventual approval by the Government of the day. These plans are reviewed from time to time. Some national parks e.g. the D'Entrecasteaux, have enclosed within their boundaries areas which are proposed national parks but not yet declared. Likewise some proposed nature reserves, though managed as nature reserves, remain to be declared. The Perup is referred to throughout the text as the Perup Nature Reserve for simplicity whereas it has yet to be declared a nature reserve.

CONTENTS

FIGURES

TABLES

APPENDICES

CHAPTER 1
INTRODUCTION - THE KARRI AND SOUTHERN FORESTS

Karri forest conservation and management can not be treated in isolation. The karri forest is not a single entity; an unbroken tract of karri from one end to the other. Karri occurs in mixtures with other forest tree species, and is patchily distributed across an area far greater in extent than its total occurrence.

The conservation and management of the karri *(Eucalyptus diversicolor)* is therefore inextricably linked with the surrounding forests, woodlands, sedgelands, swamps and other vegetation formations which together form the southern forests of Western Australia (see Appendix 7 and Fig. 1 for definition of the area).

This book deals with the southern forests in their entirety to get the true picture of conservation and management in the karri forest.

The southern forests of Western Australia are one of the most important and beautiful areas of Australia. These forests are important for two main reasons:

- the occurrence of one of the world's tallest hardwood forests, the karri forest; and
- the occurrence of a unique assemblage of animals and plants which have special conservation significance.

Compared with many great forest associations of the world, the karri forest occupies only a relatively tiny area - less than 200 000 ha - and is confined to the south-west corner of our vast continent. Nevertheless, it occupies a very special place in the minds of many Australians. There are three reasons for this:

- the forest has outstanding economic value, through timber production and as a tourist attraction;
- it has great beauty; and
- its management and protection has been a focus of political and social concern and debate for nearly 70 years.

There have been three major public controversies over the karri forest during the twentieth century.

The first occurred in the early 1920s when the Government of the day decided that karri forest should be cleared and converted into dairy farms. This led to the alienation and ringbarking of thousands of hectares of prime forest country. The scheme was vehemently opposed by foresters, but in the end foundered because of the economic and social problems experienced by the settlers and the onset of the Great Depression in the 1930s.

During the 1930s, 1940s and 1950s, the Forests Department undertook a major program to secure karri forests on Crown land as State forest, effectively blocking further alienation for agriculture. This program was fiercely opposed by farming interests and by local government, but unsuccessfully. Eventually, the bulk of the karri forest that had not been alienated during the settlement scheme of the 1920s was secured as State forest.

A new controversy emerged during the 1970s and 1980s. This took the form of opposition from some environmentalists to timber cutting in the forests. During this period, land use plans for the forests were developed and refined. The current plans were published in 1987; the new plans will be released at about the time of publication of this book. They seek to accommodate all interests and demands.

Forest utilization, management and protection slowly evolved in the context of the major controversies described above. The timber industry was established long before formal forest management. There were large sawmills at Denmark and Karridale by the 1890s, and at Deanmill, Jardee and Pemberton in the years 1912-1914. The Forests Act was passed in 1918 and the Forests Department established in 1919, but the first professional forester did not take up office in the karri forest until 1929.

Forest management has developed to a sophisticated level since the 1930s. Developments include:

- detailed type-mapping of the forest;
- an effective fire management system;
- efficient techniques for regeneration following timber cutting;
- biological surveys throughout the forests;
- intensive research into soils, vegetation associations, fauna and fire ecology and the hydrology of the forest.

Current management provides for a range of recreational pursuits and for planned protection of landscape values.

Under the 1987 management plans about half of the karri forest is managed for nature conservation and recreation, and the remainder for multiple uses, including timber production.

Nevertheless, disagreements continue about the commercial use of the karri forest for timber production. Disagreements centre on:

- the technique used in harvesting and regenerating karri and karri-marri *(E. callophylla)* stands - i.e. clearfelling - which is believed by some people to cause permanent environmental damage to the forest;
- the conversion of logs unsuitable for sawmilling to woodchips exported for paper manufacture, which is regarded by some people to be an unacceptable use of large and beautiful trees;
- the cutting of 'old growth' forest - i.e. mature, virgin forest - which is regarded by some people as a non-renewable resource; and
- the 'security of purpose' for areas of State forest zoned for conservation outside the reserve system.

All of these issues are contested and none are easily resolved. For one thing, they are charged with emotion and therefore hard to debate in an objective way. Furthermore, it is almost impossible to separate value judgements from the complex economic, social and political considerations that affect every forest policy decision.

The second reason for the importance of the southern forests is their conservation significance.

The debate on the karri forest has focused attention on only one portion, albeit a visually spectacular one, of the southern forests. This has tended to distract from many important biological values of the area. The karri forest occupies slightly less than one-sixth of the area of State forests, national parks and nature reserves in the area. Biological and conservation values are spread across the entire extent of the southern forests.

The conservation significance of the southern forests is best seen from the perspective of the geological time scale. Palaeobotanical evidence suggests that 50 million years ago during the Eocene, forests akin to the present rain forests of the tropics were once spread across much of the Australian continent. Since that time there has been a general drying of the climate, the area of the rain forest has shrunk, and there have been enormous adaptive changes among the flora and fauna. During much of this time the south-west flora and fauna have been separated from those of the rest of Australia, and adaptation has proceeded along independent lines, resulting in a unique biota. The southern forests contain a significant proportion of this unique biota.

In addition, parts of the southern forests have provided a refuge for mesic elements of the old Gondwanan flora and fauna during these long epochs of climate change. As a consequence they harbour a relictual element of the Australian biota which is to be found nowhere else.

The aim of this book is to present and interpret data about the conservation significance and the status of the karri forest that can be used in the ongoing processes of review and discussion on these difficult issues. It summarizes research and survey work carried out in the southern forests over many decades, and examines the reserve and management systems that have been developed and put in place.

CHAPTER 2
NATURAL ECOSYSTEMS OF THE SOUTHERN FORESTS

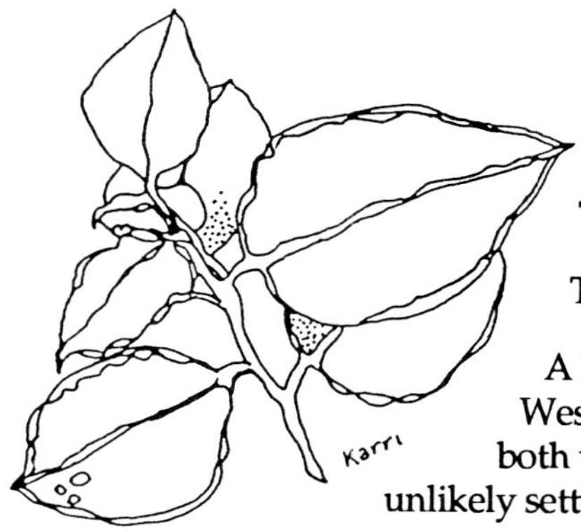

The Area and its Natural Features

The Southern Forests

A large portion of the Australian continent is arid. In Western Australia only the extreme south-western corner is both well watered and extensively forested. It is in this unlikely setting that one of the world's tallest trees, karri, grows.

Karri does not occur as a single unbroken tract of forest; its occurrence is closely related to soil moisture and other edaphic factors. Over most of its range, with the exception only of the very central portion, it occurs in mixtures and as a mosaic with other forest formations: woodlands, heathlands and sedgelands. The area referred to as the southern forests in this book includes all major occurrences of high open karri forest and associated vegetation formations.

The major karri outlier, which occurs in the Porongurup Range to the east of the southern forests, is included where relevant. This outlier is entirely within the Porongurup National Park. The southern forests boundary, as described in Appendix 7 and indicated in Figure 1, is almost identical to the area surveyed by Christensen *et al.* (1985). The southern forests include, in addition to the major karri forest occurrence, the north-western outlier of karri found to the north of Augusta (the Boranup forest), the area known as the Donnybrook Sunklands and the Perup Nature Reserve on the north-eastern boundary of the area. This area forms a natural region of predominantly high rainfall forest and adjoins the northern jarrah forest area treated by Dell *et al.* (1989) in their book *The Jarrah Forest*.

The southern forests contain some of the most extensive tracts of forested land and has some of the most spectacular scenery in the forested south-west. The landscape has great variety, from the tall stately karri and tingle *(Eucalyptus guilfoylei, E. jacksonii)* forests, to open sedgelands interposed with the huge granitic monadnocks of the south coast, to the parkland-like wandoo (*Eucalyptus wandoo*) woodlands to the north-east. The coast has a rugged grandeur with steep cliffs and rocky headlands, often flanked by shimmering white beaches and crystal clear blue-green water. At Walpole the tall forest comes to the waters' edge, while further to the north-west, vast mobile sand dunes are engulfing the forest.

Of the 2 114 700 ha encompassed by the southern forests, 1 322 400 ha, or 62.5 per cent, is State forest, national park or nature reserve, and is managed for nature and resource conservation purposes (see Table 4). More than half of the State's hardwood production comes from the forests of the south. There is also an expanding plantation program of both softwood and hardwood in the area. Tourism is an important and growing industry; the forests, parks and reserves of the area are a major recreational resource. The availability of water in the area has led to a diverse farming industry that ranges from fruit, vegetable and dairy production, to sheep and cattle grazing. Of less significance are marron and trout farming, deer farming and commercial wildflower production.

Mining in the area is more localized, comprising mainly tin mining at Greenbushes (to the north of Bridgetown), a small peat mine near Lake Muir, and mineral sands at Beenup and Jongardup. The area's economy is therefore largely related to primary industries which are an integral part of the area and an important part of the State's economy. The people of the area and

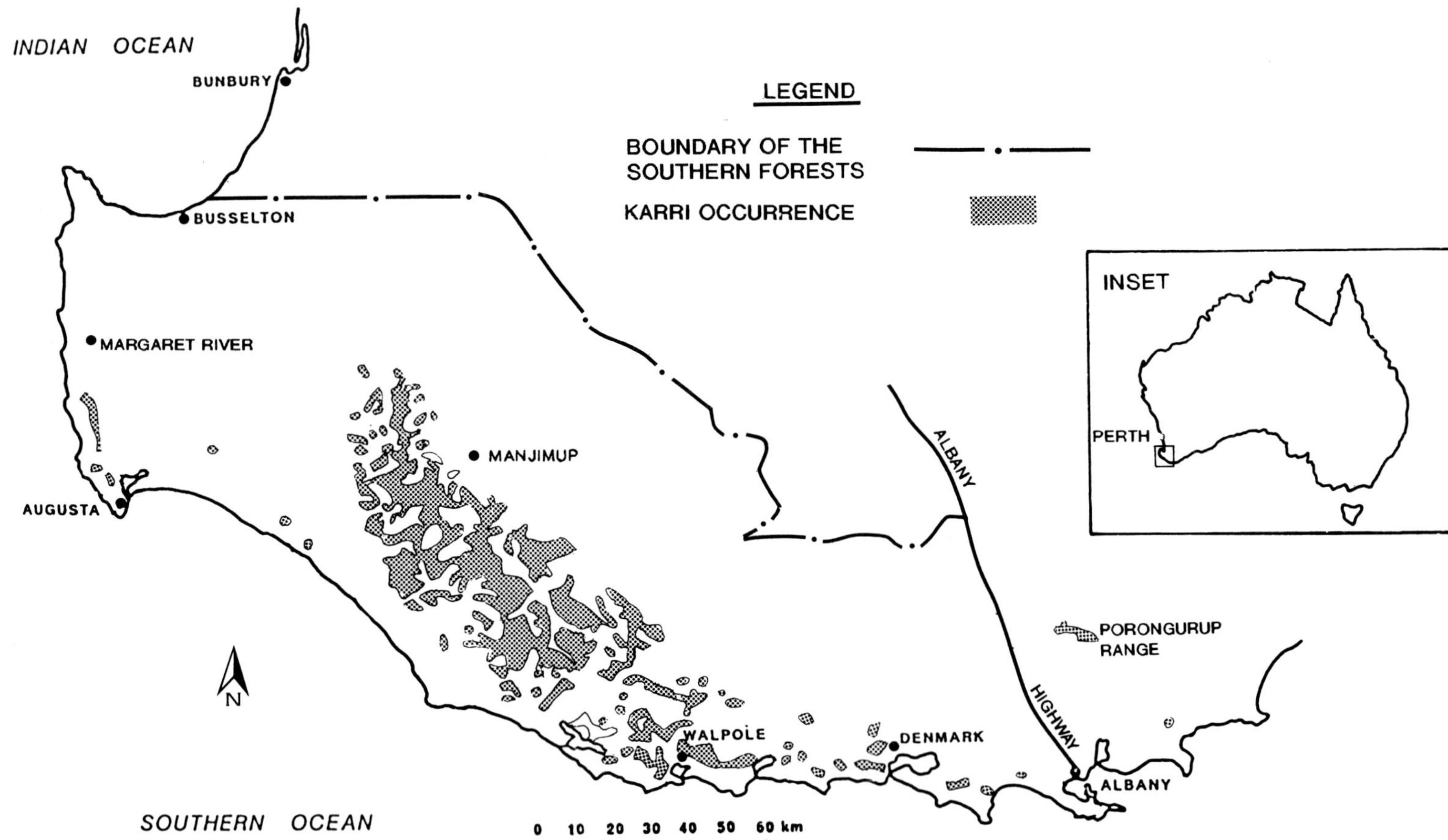

Figure 1
Karri (*Eucalyptus diversicolor*) occurrence.

their activities, many of which, such as forestry, involve the natural environment, are also a major consideration in reviewing the management of natural lands.

Geology and Physiology

The southern forests form the south-western fringe of the Great Plateau of Western Australia (Jutson 1914). It is composed of ancient Precambrian crystalline rocks partially overlain by various consolidated and unconsolidated sediments. The land surface has a long history of weathering and denudation.

The Precambrian basement is formed by rocks of Archaean and Proterozoic origin and include granitic gneisses and migmatites. This basement is partially overlain by Eocene marine sediments that are dominated by Pallingup Siltstone in the south (Cockbain 1968), but include some sandstone and limestone.

In the north-western part of the area are sedimentary rocks known as the Donnybrook Sunklands. They are possibly of Cretaceous and Permian age. This area, down-faulted between the Dunsborough and Darling Faults, is the southerly extension of the Perth Basin and is flanked in the east and the west by rocks of Precambrian age.

A weathered mantle, the lateritic profile of Stephens (1946), overlays both the Precambrian and the sedimentary rocks extensively. Differential erosion of the mantle, with exposure of the deeper horizons including the country rock, has resulted in depositions of detritus downslope.

Small areas of unconsolidated sediments, usually grey sands, occur in a variety of topographic positions. Fossils of terrestrial flora rarely occur and have been placed tentatively in the Cainozoic era (Wilde and Walker 1984). On the coast, Pleistocene Tamala Limestone (Logan 1968) and unconsolidated Holocene aeolian sands overlay earlier series. Caves, some of which contain archeological remains and fossils, are common in the limestone, especially on the Leeuwin-Naturaliste Ridge.

Within the southern forests, the Great Plateau gradually slopes towards the southern ocean and forms part of the Ravensthorpe Ramp of Cope (1975). In the north-west, the plateau is delimited by the southern extension of the Darling Scarp. Southward, the terrain is patterned by ridges of granitic rocks alternating with broad swampy corridors. Traces of Eocene marine sediments have been observed in the corridors to the north of Walpole (Darragh and Kendrick 1980). In areas around Northcliffe such as the Pingerup Plains and the Chudalup Plains, sandy and swampy tracts increase in extent. Granitic monadnocks are sometimes exposed as emergents above the sandy tracts. Mt. Chudalup (188 m) and Mt. Pingerup (182 m) are two prominent examples.

The broad ridges of Tamala Limestone are an important feature of the coastal zone. They rise to 100 m and may act as barriers behind which estuaries, such as Wilsons Inlet, have developed. A system of parabolic dunes of Holocene age extend from the coast, overlapping the broad limestone ridges, as well as the various alluvial deposits and the granitic basement rock. Much of the south coast is characterized by a succession of small bays with granitic headlands linked by barriers of Tamala Limestone, often as steep cliffs (Churchward *et al.* 1988). These bays are less common on the western coastline, so long stretches of beach are more frequent.

Eight major rivers drain the area: the Donnelly, Warren, Gardner, Shannon, Deep, Frankland, Kent and Denmark Rivers. Most streams that arise in the area have a characteristic sequence of valley-forms, from broad, flat swampy headwaters through several stages of dissection to deep channels. The major streams are sluggish in their lower courses. Within the gently sloping plateau there are clearly defined landscape units based on degree of dissection, the shape of valleys, and on the coast, position, relief, drainage status and stability. These landforms have characteristic soils which have been described by McArthur and Clifton (1975).

The deeply weathered mantle acts as an aquifer, but it also acts as a store of salts which is often in balance with the evapotranspirative capacity of the vegetation. The occurrence of salts in the weathered zone of the soils in the central portion of the area is discussed by Johnston *et al.* (1980). Soluble salts in the weathered mantle are a general feature of other parts of the Great Plateau, with higher concentration in areas of lower rainfall (e.g. Dimmock *et al.* (1974)).

The headwaters of streams rising north-east of the area are broad valleys with salt lakes that contribute naturally to the salinity of surface water (e.g. the Warren, Frankland and Kent Rivers). Saline seeps appear occasionally on other streams within the area, and are largely associated with agricultural clearing. Stream monitoring suggests that salinity is increasing in many of the major streams and rivers (Collins and Barrett 1980; Collins and Fowlie 1981). This increase is caused by permanent clearing of native vegetation for agriculture.

Landforms and Soils

The soils and landforms of the area are well known. Several soil surveys have been carried out in different parts of the area: Hoskings and Burvill (1938) at Denmark, and Smith (1951a, 1951b) in the Lower Blackwood and Margaret River districts and in the Frankland River valley. McArthur and Clifton (1975) and Churchward *et al.* (1988) carried out landform and soils surveys of the Pemberton and south coastal areas respectively.

Churchward *et al.* (1988) recognize 35 landform and soil units in the southern part of the area. These units are grouped according to whether they are associated with granitic rocks or sedimentary formations, some being associated with extensive areas of unconsolidated superficial deposits. The groupings are as follows:

- Units developed on granitic rocks and associated unconsolidated sediments.

  (a) Plateau Elements
  (b) Hills and Hilly Terrain
  (c) Swampy Terrain

- Units developed on siltstones and sandstones.

  (a) Plateau Elements
  (b) Swampy Terrain

- Units developed in coastal aeolian and fluviatile sediments.

  (a) Dune systems
  (b) Coastal Swampy Terrain

- Units associated with drainage lines.

  (a) Major valleys
  (b) Minor valleys

These units are all described in terms of physiology, geology, soil morphology and associated native vegetation.

McArthur and Clifton (1975) described the soils according to Northcote (1971). They recognized six major groups that are represented throughout the area.

1. *Laterites and Ironstones*

The laterites and ironstones that are typical of the Darling Scarp occur with less frequency in the southern forests. Mostly present in the northern sectors of the southern forests area, they are characterized by a layer of light brown sand and sandy loam dominated by ferruginous gravel or blocks of duricrust; yellow mottled clay usually occurs within the first metre. Laterites are generally found high in the landscape where there is less erosion. Forests on these soils are typically jarrah (*Eucalyptus marginata*) or jarrah-marri (*E. calophylla*) mixtures with *Banksia grandis* as a common understorey species.

2. *Red Earths*

These younger soils typically develop in a dissected landscape. They consist of a reddish-brown loam or a sandy loam surface which changes gradually to a red clay at 50 cm depth. Red earths support high open forests of karri, or where rainfall is insufficient, an open forest of yarri (*E. patens*) or jarrah.

These soils occur when basic or acidic rock has been exposed by stream truncation, which removes the laterite formation and allows new soils to develop.

3. *Podzols*

Podzolic soils occur in the middle and lower landscape positions and appear to be formed in a dissected lateritic landscape. There are several distinct types of podzols:

(a) *Red Podzols:* These soils have surface layers of red-brown sandy loam with an abrupt change to a red clay horizon at 40 cm depth. Within the area, they generally support jarrah or jarrah-marri open forest, and in wetter areas high open forest of karri-marri.

(b) *Yellow Podzols:* The common features of these soils is the distinct change in texture from the topsoil to a yellow mottled clay subsoil. They cover a wide range of sites in the southern forests. Yellow podzols appear to have developed from the exposed mottled zone of the laterites, and are thus widespread in areas of intermediate relief.

The yellow podzolic soils support a wide range of vegetation associations, from treeless flats in the swampy drainage lines to high open forest of marri-karri in the most favourable situations, and marri-jarrah and wandoo in drier eastern areas.

(c) A minor form of podzol has developed on certain swampy upland sites. These soils have a dark grey peaty sand surface and a plain mottled clay subsoil. Water-rounded quartz stone is often associated with this form of podzol which suggests that they consist partly of transported materials. They are often underlain by quartzite.

4. *Undifferentiated Sands*

Sandy soils often develop downslope from the lateritic ridges. The colour of these soils may vary from pale yellow through to reddish-brown, and ironstone gravel may be present in the subsoil. These sands generally support low woodlands of *Banksia* spp. or *Allocasuarina fraseriana*. A grey leached form of these sands occurs in broad drainage lines where it often supports a sedgeland community.

5. *Calcareous Sands*

Along the coast, southwards from Augusta, a zone of soils derived from calcareous beach sands occurs in the form of consolidated sand dunes. These sands support coastal scrubland, and open woodland of the Western Australian peppermint (*Agonis flexuosa*) and yate (*E. cornuta*).

6. *Alluvial Soils*

Alluvial soils, a minor soil type, occur as narrow bands along major streams. They are extremely variable, but have a medium texture throughout the soil profile, high organic content, and are dark brown or dark grey.

A further soil type recognized by Churchward *et al.* (1988) consists of undifferentiated siliceous sands. This minor soil type occurs where drainage is not restricted. These soils generally have no horizon development other than organic accumulation at the surface.

Climate

The southern forests have a mild mediterranean climate with warm summers, mild winters, and only occasional frost. During the winter months, mean temperatures in the area are dependent on distance from the sea; isotherms are generally parallel to the coast, and temperatures decrease with increasing distance from the sea. In summer, mean temperature is more affected by latitude than proximity to the coast. The growing season ranges from 7.5 months at Manjimup to 10 months in the extreme south (see Bureau of Meteorology 1965 and 1975).

Seasons are distinct with most of the rain falling in the 'winter' (May to October). Limited summer rainfall may be experienced along the south coast, but summers are generally dry.

Rainfall isohyets are approximately parallel to the coastline, ranging from more than 1400 mm in the south-west to less than 635 mm in the north-east of the area: the point furthest from the coast (Fig. 2a). An anomaly in the rainfall pattern occurs at Margaret River, where there is an isolated area of higher rainfall. This coincides with a major karri outlier.

Forests and Plant Formations

The Distribution of Major Forest Formations in relation to climate, landforms and soils in the southern forests

Karri forest, more because of its spectacular beauty than its overall extent, dominates the southern forests. It occurs in the extreme south-west of the area south of a line from Nannup in the north-west through Manjimup to the Frankland River in the south-east. It extends further eastwards to Denmark, Torbay and Albany, but in this area it is confined to within 15 to 20 km of the coast. Two main outliers occur, one on soils derived from coastal limestone at Karridale (Hall *et al.* 1975), the other on soils derived from weathered granite on the Porongurup Range north of Albany. Smaller outliers occur at Yallingup, Margaret River, Black Point, Rocky Gully and near the south coast adjacent to Mt. Manypeaks, east of Albany (Fig. 1). Karri is largely confined within the 1100 mm isohyet (Fig. 2a). Where it occurs outside this limit, e.g. Rocky Gully, Porongurup Range and Mt. Manypeaks, groundwater is supplemented by runoff. There is evidence of a wider distribution in the past in times of greater rainfall (Churchill 1968).

Karri grows in a variety of soils developed over granite, granite-gneiss, limestone, laterites and sand (McArthur and Clifton 1975). Karri attains its best development as a tree on red earths in the high rainfall (>1250 mm per annum) central parts of its range in the vicinity of Pemberton.

The distribution of karri within its main range is closely related to the presence of younger red earths, and red-brown to yellowish-brown podzolic loams whose occurrence relates to topography (McArthur and Clifton 1975; Bradshaw and Lush 1981). In the north-western parts of its range, in the Donnelly River valley, karri is typically found in incised valleys. Further up-slope it occurs in mixture with marri, and on the ridges it disappears, being replaced by jarrah. Further to the south, particularly in the catchment of the Warren River, the old laterite duricrust

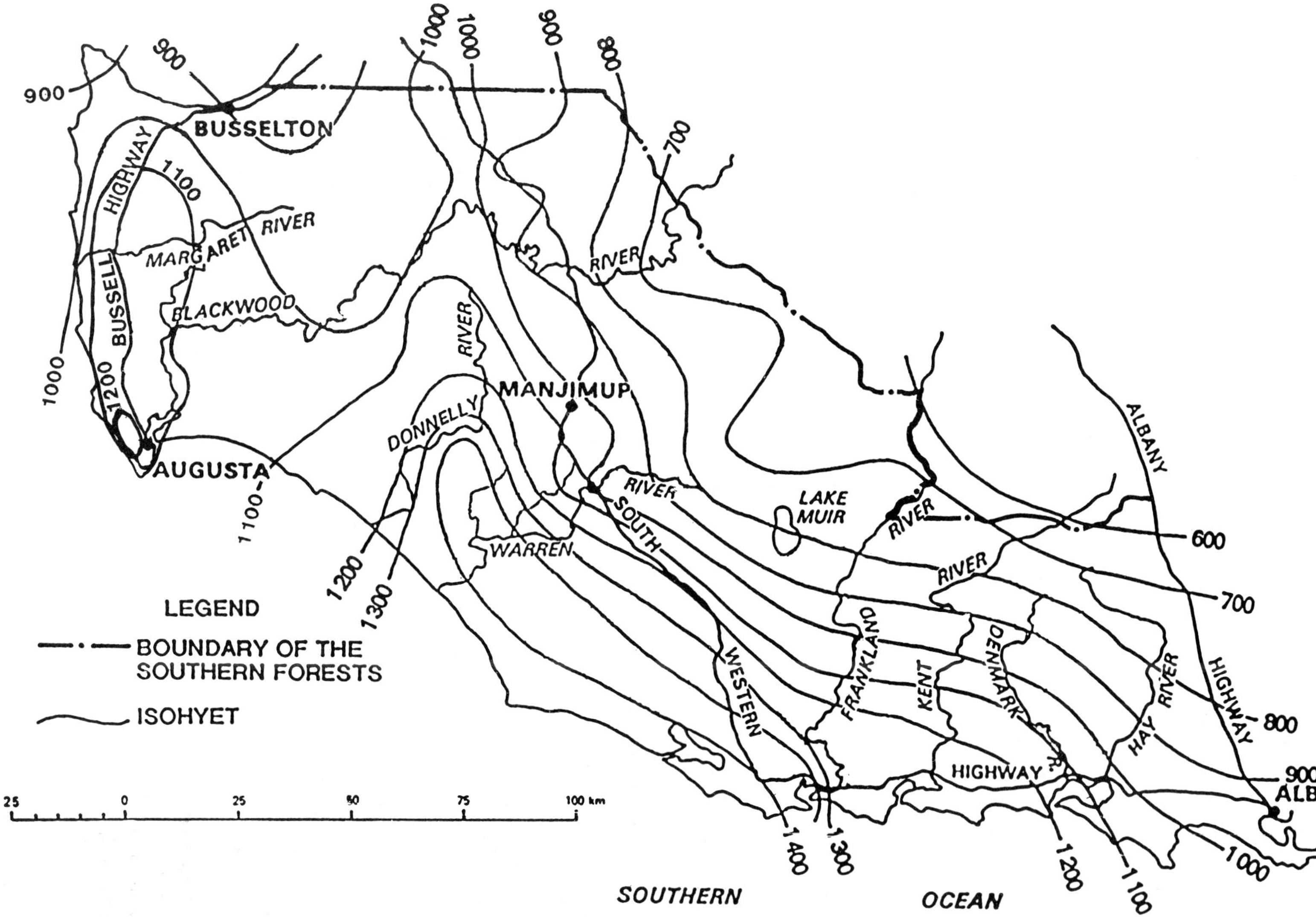

Figure 2a
Rainfall distribution in the southern forests.

(McArthur and Clifton 1975) becomes more dissected and the rainfall is greater, and small tributaries to the major river develop numerous shallow valleys. Here, karri extends from the valley bottom to the mid and upper slopes. Here the largest contiguous area of karri forest occurs.

Further to the south the land was once subject to inundation by the sea and there are now widespread Quaternary deposits of sandy soils, forming extensive seasonally inundated flats, supporting scrub and sedgeland communities (McArthur and Clifton 1975). Remnants of the old land surface protrude as monadnocks emergent from the broad flat sandy plains and swamps. In this landscape karri occurs in the red earths that have developed around and below the granitic outcrops. It is also found along the valleys of rivers, which have dissected these sandy flats and exposed the red granitic gneissic soils beneath.

A similar situation is found in the north-eastern section of the main karri range where depositional soils fill many of the ancient river valleys (Bradshaw and Lush 1981). Extensive forests of karri and marri mixed with yellow tingle (*E.guilfoylei*) and red tingle (*E.jacksonii*) occur on yellow and red duplex soils derived from the old land surface which outcrops along the south coast near Walpole and Denmark. Elsewhere, on weathered and less fertile soils within the 1100 mm isoyhet, jarrah or jarrah-marri forests predominate. Jarrah is more prevalent on the lateritic soils and ironstone, whereas marri is more common on sandier soils. Poorly drained sites and sands along the coast support a range of woodlands, heathlands and sedgelands.

Those parts of the area with an annual rainfall below 1100 mm, such as the Donnybrook Sunklands to the north-east, the area to the north and east of Manjimup, and the Frankland River area, also support jarrah and marri forest. Small areas of yarri occur on the better soil where the rainfall is 900 to 1000 mm. Limited areas of wandoo *(E. wandoo)* forest and woodlands occur on the clay soils in the north-eastern sector of the area where the rainfall is below 900 mm.

Vegetation Systems and Plant Formations and Associations

The wide range of soils, together with climatic gradients and to a lesser extent topographical differences, are reflected in the high number of plant formations and associations that occur within the southern forests. Smith (1972, 1973) recognizes eight vegetation systems in the area to the west of the Frankland River, based on the soil systems of Smith (1951b) and McArthur and Bettenay (1960).

These systems are: The Chapman, Boranup, Pinjarra Plains, Spearwood, Scott River, Nornalup, Darling and the Jingalup Systems. Two of these support karri forest: Nornalup, which encompasses the main karri zone, and Boranup, which contains a few stands of karri on stabilized sand dunes to the west and south of the Nornalup system.

Smith (1972, 1973), has further described and mapped the major vegetation formations of the area using structural criteria of the tallest stratum, together with projected foliage cover (the system of Specht 1970). Major plant associations within each formation were identified using species composition.

A total of 19 forest, woodland, scrub, heath and herbland formations were recognized encompassing some 51 plant associations. To this variety can be added the further formations and plant associations recognized by Beard (1980) in the area to the east of the Frankland River. Because of slight differences in mapping methodologies of the two authors, the true totals are not easy to define accurately.

A simplified description of the major formations and plant associations within the southern forests is provided in Christensen *et al.* (1985). (see also Fig. 2b.) These authors briefly described the range of plant association that occur under seven major groupings: high open forest, open

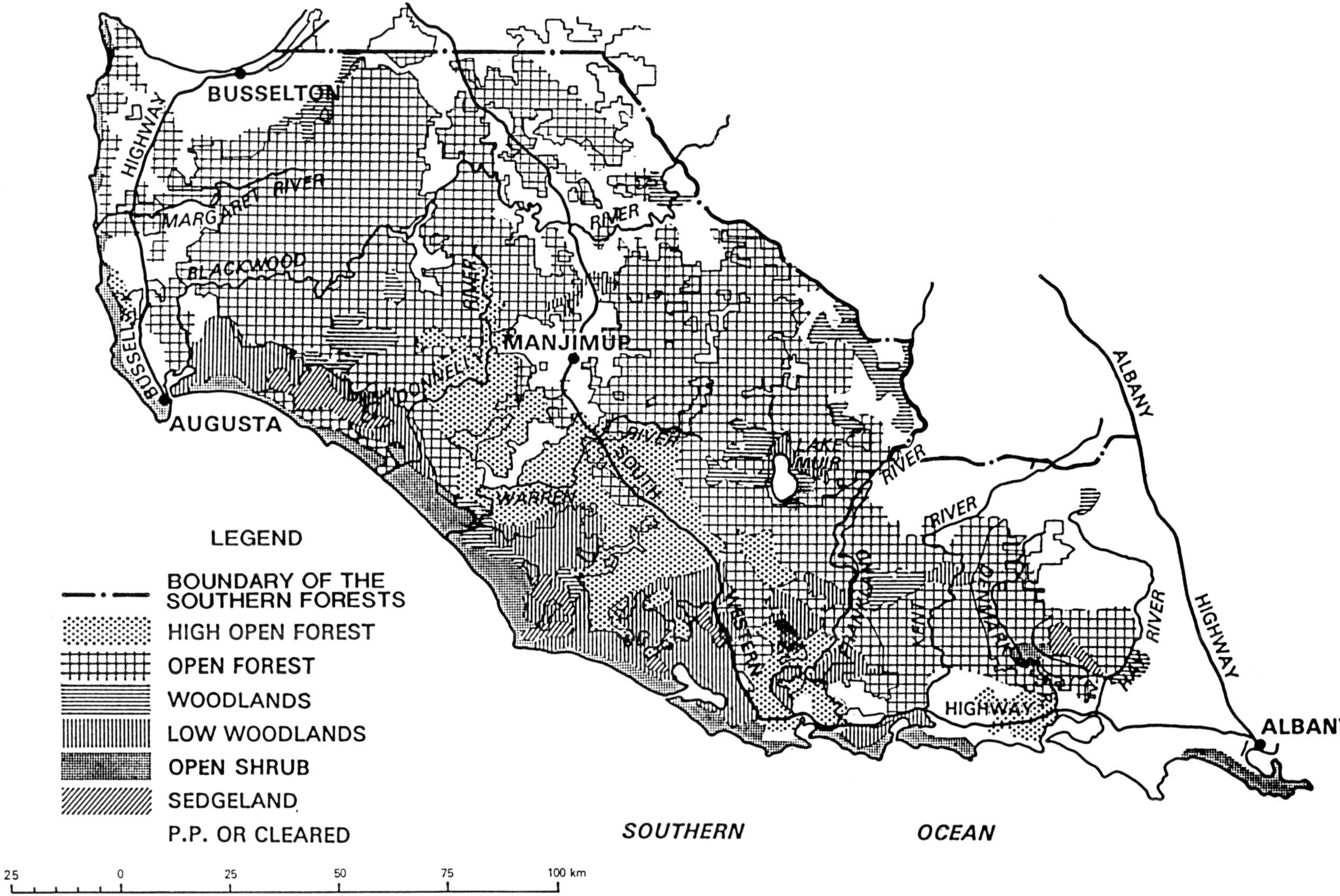

Figure 2b
Major vegetation formations of the southern forests.

forest, woodland, low woodland, open and closed heathland and sedgeland. In addition, they recognized two distinctive habitats: granitic monadnocks, and waterways and wetlands.

The following description of the major plant formations of the southern forests is taken from Christensen *et al.* (1985).

1. *High Open Forest*

High open forests are typically karri (Plate 1) forest, or karri and marri forests, less frequently mixed with jarrah and yarri. Near Walpole, yellow tingle, red tingle (Plate 2) or Rate's tingle (*E. brevistylis*) may be present. The main understorey tree species are karri oak (*Allocasuarina decussata*) and Western Australian peppermint. Common understorey species are *Acacia urophylla, Trymalium floribundum, Paraserianthes lophantha, Pimelea clavata, Thomasia quercifolia, Chorilaena quercifolia* and *Agonis parviceps*. In the south-easterly sections of the area, *Acacia pentadenia, Lepidosperma tetraquetrum* and *Lepidosperma effusum* are the dominant understorey species, while in the northerly sections of the area, in the Donnelly River valley, *Bossiaea laidlawiana* is more common.

High open forests have developed on the most fertile soils within the area where annual rainfall is greater than 1100 mm.

2. *Open Forest*

The open forest is generally composed of jarrah (Plate 3) or jarrah-marri mixtures, with yarri, flooded gum (*E. rudis*) and yate occasionally occurring in small patches by themselves. Common understorey species that may be present throughout the area include *Banksia grandis, Allocasuarina fraseriana, Persoonia longifolia* and *P. elliptica*. On grey sands, *Banksia attenuata* and *Xylomelum occidentale* may occur, and on damper sites *Agonis flexuosa* may be present (Plate 4). Common shrub species are *Bossiaea linophylla, B. ornata, Macrozamia riedlei, Hakea amplexicaulis, Xanthorrhoea preissii* and *X. gracilis, Acacia pulchella* and *Agonis parviceps*. On the drier sections, *Hakea lissocarpha, H. undulatum* and *Trymalium ledifolium* are common species.

Open forests grow on lateritic and podzolic soils, and in drier regions.

3. *Woodlands*

Woodlands are most common in the more easterly sections of the area, and are mainly composed of wandoo (Plate 5) or jarrah or marri or a mixture of these species. Both yate and flat-topped yate (*E. occidentalis*) occur as woodlands on suitable sites, and marri may be present on the sandier soils in the low lying areas. On very sandy soils in the higher rainfall areas, *Allocasuarina* spp. can form woodlands, frequently in association with stunted jarrah or *Banksia* spp. In the south-east of the area, Albany blackbutt (*E. staeri*) grows as a woodland either alone or in association with *Allocasuarina fraseriana* and *Banksia attenuata*. A notable woodland formation occurring in the Nornalup area is that of the red flowering gum (*E. ficifolia*), which is confined to this locality, and only grows on consolidated dunes or, away from the coast, on sandy gravels.

The scrub layer in the wandoo woodland is frequently sparse and quite variable in species' composition. In the jarrah and marri woodlands, the understorey is less open and clumped, with species such as *Bossiaea ornata* and *Hakea lissocarpha* being common. Flat-topped yate generally has an understorey of sedges, as this habitat is seasonally waterlogged. In the casuarina woodlands, the understorey is dense, with species such as *Anarthria scabra* and *Beaufortia decussata* being prominent. Where Albany blackbutt occurs with *Allocasuarina* spp., *Beaufortia anisandra* is usually present in the scrub layer.

4. *Low Woodlands*

These vegetation types, which occur throughout the area, vary greatly in both species' composition and soil types.

Banksias are a common component of the low woodlands. On dry sandy sites, *Banksia attenuata*, *B. ilicifolia*, or *B. grandis* are generally present. On wetter sites, such as peaty sands, these species are often replaced by *B. littoralis* and *B. quercifolia*, frequently in association with the Western Australian Christmas tree (*Nuytsia floribunda*) or *Melalueca* spp. On some wet sites, *Melaleuca* spp. may form low woodlands, with common species being *M. preissiana*, *M. rhaphiophylla* and *M. cuticularis* (Plate 6). Near the coast on stabilized sand dunes, *Agonis flexuosa* is the major species, although it may associate with several *Melaleuca* spp. On extreme sites generally constituted on very shallow soils over rock, clay or consolidated dunes, jarrah, marri, yate, bullich (*E. megacarpa*) and flooded gum may all form low woodlands.

In the Whicher Range in the north-west extremity of the area, woodlands are primarily of mountain marri (*E. haematoxylon*), the distribution of which is restricted to the ranges (on sandy lateritic gravels) and to a few localities further north along the Darling Scarp.

5. *Closed Scrub - heath*

This formation is generally associated with permanently moist sites characterized by such species as *Melaleuca* and *Agonis*, *Kunzea ericifolia*, *Banksia quercifolia* and *Homalospermum firmum*. On the calcareous dunes near Karridale, the mallee *E. calcicola*, together with a mallee form of bullich and yate, form patches of closed scrub in association with *Melaleuca huegelii*, *Diplolaena dampieri*, *Scaevola nitida* and *Acacia decipiens*.

6. *Open Scrub - heath*

Limited areas of the mallee eucalypt *E. anceps* and *E. decipiens* occur in some of the eastern sections of the area. Scattered patches of jarrah, marri and yarri grow on 'islands' a few centimetres higher than the surrounding sedgelands. These species occur in seasonally wet areas along the coast eastward from Augusta. In some localities, notably the Donnybrook Sunklands and the catchments of the Kent, Frankland and Hay Rivers, large open areas with scattered stunted jarrah and marri, *Kingia australis* and *Allocasuarina humilis* occur on very shallow clayey soils.

7. *Sedgelands*

Sedgelands (Plate 7) composed of mixed monocot species are found on the upper reaches of the Margaret River, around Lake Muir and associated lakes, and on some of the large peaty sub-coastal flats, particularly common to the south, where drainage is impeded. Consequently, sedgelands are under water during the winter months. Genera typically found associated with sedgelands include the following: *Evandra*, *Lepidosperma*, *Gahnia*, *Schoenus*, *Leptocarpus*, *Loxocarya*, *Restio*, *Juncus and Isolepis*.

8. *Granitic Monadnocks*

A number of granitic monadnocks, consisting of huge granitic bosses and boulders, occur in the southern part of the area (Plate 8). The vegetation on these granitic outcrops is different from that of the surrounding forest, often comprising large numbers of endemic species. Species commonly found in association with these rocks include *E. megacarpa*, *Lepidosperma effusum* and a variety of ferns, mosses and lichens.

9. *Waterways and Wetlands*

Waterways and wetlands form a variety of specialized, permanently moist habitats, ranging from reed swamps (*Cladium* spp.) and paperbark swamps (*Melaleuca* spp.) to riverine habitats dominated by *Agonis juniperina, Oxylobium lanceolatum* and *Banksia* spp. (Plates 9 and 10).

Several of the vegetation associations are unique to the area, including the karri and tingle forests, which are the only extensive occurrences of high open forests in Western Australia, the woodlands of red flowering gum, and several of the heath and sedgeland associations. Karri oak is largely restricted to the area and the woodlands of Albany blackbutt are some of the few remaining protected woodland areas of this species. Only the open forests of jarrah and marri and some of the woodland formations such as wandoo and flat-topped yate are common outside the area.

The understorey communities of the high open karri and the open jarrah and marri forests have been studied in detail. Thirteen plant communities have been recognized within the high open karri forest. The distribution of these is determined largely by climatic factors, in particular precipitation in the wettest quarter and radiation in the driest quarter, and to a lesser extent edaphic factors, especially soil pH and phosphorus content (Inions *et al.* 1990a and 1990b). The main high open karri forest occurrence may be subdivided into three zones on the basis of climate:

- a north-western and central zone characterized by high rainfall markedly seasonal in its distribution, and high summer temperatures;
- a north-eastern and eastern zone with comparatively lower rainfall and higher temperatures; and
- a south-coastal zone where rainfall is less seasonal, there is some precipitation in summer, and temperatures are lower.

Many of the 13 plant communities are centred on one or the other of these climatic zones; a few are restricted to one zone only. The occurrence of each of the plant communities within each climatic zone is determined largely by edaphic factors (Inions *et al.* 1990a and 1990b).

Strelein (1988), using plant indicator species and ordination techniques, recognized 17 site-vegetation types in the southern jarrah forest. McCutcheon (1978 and 1980) recognized seven site-vegetation types in the jarrah forest of the Donnybrook Sunklands.

Three studies have examined ecotones in addition to pure forest stands. Wardell-Johnson *et al.* (1989) derived a floristic classification of the Walpole-Nornalup National Park using clustering and ordination techniques. They recognized 12 community types which have association with the soil and landform units of Churchwood *et al.* (1988).

Site vegetation types have also been studied in relation to animal distribution in the jarrah and wandoo forests of the north-eastern sector of the area (see Christensen 1980). A smaller study by Bridgewater (1981) used vegetation description and classification to define the karri forest boundary near Pemberton.

The different forest and vegetation associations of the southern forests are thus comparatively well known. In comparison with the northern jarrah forest (see Havel 1975a, 1975b and 1989), an area of comparable size, the forest and vegetation associations of the southern forests are more complex and have greater diversity, although the karri forest *per se* may be the most species-poor (Wardell-Johnson *et al.* 1989).

Plant and Animal Species

Flora

Vascular Plants

Although no comprehensive flora has been published for the southern forests, information on the distribution and composition of the flora of the area is contained in a series of general works on the vascular flora of Western Australia, e.g. Blackall and Grieve (1954-1988), Erickson *et al.* (1973), Beard (1981). Green (1985) lists the whole vascular flora of Western Australia. There are a number of flora lists, more or less complete for specific and defined areas within the area. On the south coast, lists are available for the Torndirrup, West Cape Howe and William Bay National Parks near Albany (Keighery, in prep (a)), and the Walpole-Nornalup National Park (Wardell-Johnson in prep). There is also a checklist of plants for the Porongurup Range National Park (Abbott 1980; Keighery, in prep (b)).

Smaller lists exist for the Lefroy and Four Mile Brook (Dames and Moore 1982) and Big Brook (Dames and Moore 1985) areas in the karri forest near Pemberton and the Smiths Brook Nature Reserve (Griffin 1985). Christensen *et al.* (1985) also published a list of some 900 species, mainly of woody flora, for the area, and rare and poorly known flora have been surveyed in the Leeuwin-Naturaliste National Park (Frewer *et al.* 1987).

The most comprehensive flora list for the area is of the known vascular flora of the Warren Botanical Subdistrict, which covers a large portion of the southern forests (see Hopper *et al.* 1992). This list was compiled from a search of recent Australian taxonomic literature (1950-1988) and unpublished works of Western Australian botanists. The authors also inspected specimens at the Western Australian Herbarium, the Kings Park Herbarium and the Department of Conservation and Land Management (CALM) Herbarium at Manjimup. The list contains 1771 species, including 1472 native and 299 introduced species. A further 96 native species have been added from the list of Christensen *et al.* (1985) to form the most up-to-date list of vascular flora of the southern forests (Appendix 1).

The most important families listed by Hopper *et al.* (1992) include Fabaceae (151 spp.), Orchidaceae (141 spp.), Myrtaceae (110 spp.), Proteaceae (100 spp.), and Poaceae (97 spp.). The largest genera consist of *Acacia* (68 spp.), *Stylidium* (52 spp.), *Caladenia* (51 spp.), *Leucopogon* (36 spp.), *Eucalyptus* (24 spp.) and *Drosera* (21 spp.). The large numbers of species in the families Orchidaceae, Stylidiaceae and Epacridaceae are noteworthy when compared with the State's flora as a whole (Green 1985).

The phytogeography of very few groups of vascular plants has been examined. Keighery (1984a and 1984b) concludes for the Fabaceae (Papilionaceae) that generic endemism is low for a tropical genus, but specific endemism is high in southern Western Australia. Several genera reach their highest diversity in the high rainfall forest zones, especially in the southern coastal heathlands, e.g. *Hovea, Jacksonia, Sphaerolobium, Eutaxia, Latrobea, Pultenaea,* and *Aotus*.

Among the monocotyledons, the families Restionaceae, Iridaceae, Juncaginaceae, Juncaceae, Xyridaceae, Hypoxidaceae, Centrolepidaceae and Poaceae are species rich in the high rainfall zone of southern Western Australia. This differs from the pattern shown for many woody genera which reach greatest diversity in the intermediate rainfall kwongan area of the south-west (Hopper 1979).

Hopper *et al.* (1992) list 84 species endemic to the Warren Botanical Subdistrict. They consider this total conservative since several groups, e.g. the orchids, have many endemics among undescribed taxa. Endemics are mainly confined to sedgelands, coastal heaths and granitic outcrops with the majority being found in the wetter areas between Denmark and Northcliffe. Some endemics occur on the Leeuwin-Naturaliste Ridge (9 spp.) and two are known from the Scott River plains.

Many species, including karri, are confined to high rainfall areas in damp habitats and extend outside the Warren Botanical Subdistrict only in small areas of favourable habitat e.g. karri on Mt. Manypeaks and Porongurup Range, Albany pitcher plant (*Cephalotus follicularis*) at Cape Riche and Yallingup. Hopper *et al.* (1992) consider that there are at least 52 of these 'near endemics'. They therefore consider the Warren Botanical Subdistrict a major centre of endemism for wet country species. Another feature of the flora mentioned by these authors is the large number of species that are near the end of their geographical range in the Warren Botanical Subdistrict, e.g. 56 south coast heathland species reach their western limit between Albany and Denmark.

Nineteen species of Declared Rare Flora (*Government Gazette* of 17 May 1991) occur within the southern forests (Table 1). A list of plant species of particular conservation concern, taken from CALM's list of Reserve Flora of 14 December 1990, is presented in Table 2a.

Table 1

DECLARED RARE FLORA IN THE SOUTHERN FORESTS (17 May 1991)

SPECIES	HABITAT	LOCATION
Adenanthos cunninghamii	coastal dunes	Two Peoples Bay Nature Reserve
Asplenium obtusatum	granitic outcrops	On offshore islands
Baeckea arbuscula	swamps	Millbrook Nature Reserve
Banksia goodii	jarrah low forest	In reserves outside the southern forests
Banksia verticillata	granitic outcrops	D'Entrecasteaux National Park
Caladenia 'excelsa'	Banksia low woodlands	Leeuwin-Naturaliste National Park
Caladenia 'harringtoniae'	swamps	Beedalup National Park
Caladenia huegelii	jarrah forest	Leeuwin-Naturaliste National Park
Darwinia 'ferricola'	lateritic heath	Scott River
Diuris drummondii	swamps	Lake Muir National Park
Drakaea 'micrantha'	jarrah forest	Leeuwin-Naturaliste National Park
Grevillea cirsiifolia	jarrah forest	Jarrahdale, Denmark and Dwellingup (State forest)
Isopogon unicatus	jarrah low forest	Mt. Wilyung, Torndirrup
Kennedia glabrata	granitic outcrops	D'Entrecasteaux National Park
Kennedia macrophylla	granitic outcrops	Shire reserve
Lambertia orbifolia	jarrah forest	Not in reserves
Laxmannia jamesii	swamps	Reserve near Narricup
Microtis globula	swamps	Walpole-Nornalup National Park
Verticordia fimbrilepis	jarrah forest	Kent River (State forest)

' ' - Species not yet officially named.

Table 2a

PLANTS OF PARTICULAR CONSERVATION CONCERN
IN THE SOUTHERN FORESTS

[From Department of Conservation and Land Management Priority Flora List 14 February 1990]

Priority 1. Poorly known taxa which are known from one or a few populations which are under threat.

Actinotus 'laxa'	Walpole
Andersonia auriculata	Quarram, Bow River west of Denmark
Caladenia uliginosa subsp. *'patulens'*	Harvey, Nannup
Caladenia 'winfieldii'	Strachan
Calothamnus pallidifolius	Whicher Range
Dryandra sp. *30*	Ruabon, Tutunup
Eucalyptus lane-poolei var (Whicher)	Whicher Range
Gonocarpus simplex	Cape Le Grande
Hemigenia podalyrina	Shannon Rock
Jacksonia mollissima	Bunbury-Margaret River
Leptomeria ericoides	Willyung Hill, Cowaramup
Leucopogon alternifolius	Scott River, Albany
Leucopogon polystachyus	Manjimup, Nornalup
Melaleuca basicephala	Scott River, Jondardup
Pultenaea radiata	Busselton
Schoenus acuminatus	Albany
Schoenus efoliatus	Albany
Schoenus multiglumis	Albany, Bow Bridge, Northcliffe
Sollya drummondii	Denmark
Stylidium rhipidium	Rocky Gully, Williams, Collie
Thysanotus formosus	Nannup
Thysanotus isantherus	Albany

' ' - Species not yet officially named.

Table 2a *(continued)*

Priority 2. Poorly known taxa which are known from one or few populations, at least some of which are believed not to be under immediate threat.

Acacia tayloriana	Sues Road east of Margaret River
Amperea volubilis	Walpole, Albany
Banksia occidentalis subsp. *formosa*	Black Point, Port Hughes
Boronia crassipes	Albany
Borya longiscapa	Mt. Lindesay
Caladenia 'abbreviata'	Lake Jasper, Cosy Corner
Dampiera orchardii	Tone River, Oldfield River, Ravensthorpe, Lake King
Dryandra serra	Bow River, Mt. Manypeaks
Eriochilus 'pulchellus'	Mt. Chudalup, Mt. Burnside
Grevillea fistulosa	Fitzgerald River-?southern forests
Grevillea manglesioides subsp. *papillosa*	Scott River, Cane Break
Grevillea sp.(Mt. Lindesay)	Mt. Lindesay
Hemigenia glabrescens	Albany, Scott River plain
Hybanthus volubilis	Margaret River (private property), Scott National Park
Jansonia formosa	Walpole, Margaret River, Scott River
Leucopogon bracteolaris	Stirling Range, Albany
Leucopogon multiflorus	Mt. Arid, Mt. Le Grande
Lomandra ordii	Northcliffe, Walpole
Lysinema lasianthum	Porongurup Range, Albany
Mitrasacme sp. (south-west)	Woolbernup Hill, Walpole, Capel
Pithocarpa melanostigma	Donnybrook Sunklands (very common) Albany, Millbrook
Restio gracilior	South of Busselton, Yoongrillup, Scott River
Stylidium barleei	Whicher Nature Reserve Busselton area
Thomasia brachystachys	South Stirling Range, Walpole
Tripterococcus sp. (central forest)	Scott River, Walpole, Margaret River
Xanthosia singuliflora	Mt. Manypeaks, Margaret River, King River, Stirlings

' ' - Species not yet officially named.

Table 2a *(continued)*

Priority 3. Poorly known taxa which are known from several populations at least some of which are believed not to be under immediate threat.

Acacia inops	Margaret River, Yelverton
Acacia luteola	Albany, Mt. Barker, Stirlings Ranges, Sheepwash Creek
Aotis carinata	Scott River plain
Boronia virgata	Walpole, Denmark
Cassytha micrantha	Esperance, Badgebup
Chamelaucium forrestii	Two Peoples Bay, Northcliffe
Crytandra miliaris	Manjimup, Peak Charles, Mt. Gibson, Mt. Deans
Drosera omissa	Augusta, Busselton
Eucalyptus brevistylis	Mt. Frankland, Nornalup, Walpole
Eucalyptus rudis subsp. *nov.*	Yallingup, Eagle Bay (Yallingup)
Hibbertia silvestris	Waroona, Collie, Manjimup
Hypocalymna sp. (Scott River)	Scott River, Northcliffe
Gastrolobium brownii	Walpole, Albany
Philydrella drummondii	Bayswater, Guilford, Chidlow, Helena Valley, Pinjarra, Palgarup
Pultenaea pinifolia	Busselton, Karridale (State forest)
Thomasia discolor	Albany area
Thomasia solonacea	Albany, Two Peoples Bay
Villarsia lasiosperma	Busselton, Scott River, Harvey Northcliffe, Albany, Esperance
Villarsia submersa	Lake Muir, Denmark and others

Priority 4. Taxa presumed extinct. Taxa which have not been collected or otherwise verified over the past 50 years.

Acacia prismifolia	Albany, Stirling Range
Meziella trifida	Albany
Scaevola attenuata	Albany, Cape Naturaliste
Tetratheca elliptica	Bow River

' ' - Species not yet officially named.

Table 2a *(continued)*

Priority 5. Rare Taxa. Taxa for high priority monitoring. (Species which could become rare if circumstances change)

Adenanthos detmoldii	Scott River, Blackwood River
Banksia meisneri. var *ascendens*	Ruabon, Gingilup Swamp, Scott River
Brachysema subcordatum	Porongurup Range
Caladenia 'arrecta'	Margaret River, Bindoon, Gibson
Caladenia 'interjacens'	Crystal Springs, West Cliff Point
Caladenia plicata	jarrah forest
Caladenia 'speciosa'	Ludlow, Myalup, Eaton, Yarloop, Gingin
Calothamnus graniticus subsp. *graniticus*	Sugarloaf Rock, Meelup, Eagle Bay
Chamaexeros sp. (Walpole)	Walpole-Nornalup National Park
Eucalyptus calcicola	Leeuwin-Naturaliste National Park
Franklandia triaristata	Capel, Tutanup, Jarrahwood
Grevillea ripicola	Muirillup, Collie, Kirup
Lambertia rariflora	Margaret River, Jarrahwood
Microtis pulchella	Walpole-Nornalup National Park, Northcliffe, Albany
Prasophyllum triangulare	Albany, Margaret River, Hyden
Pultenaea skinneri	Binningup, Collie, Boyanup
Restio ustulatus	Busselton, Scott River (State forest), Kent River

' ' - Species not yet officially named.

The priority plant taxa list contains 1504 species for the whole State of Western Australia (Table 2b). While the floras of some areas, e.g. the Gascoyne and the Pilbara in the north of the State, are poorly known, the floras of the areas in the south-west, including the forested areas, are relatively well known. Inspection of the Table suggests that the majority of rare and threatened taxa occur in the south-west areas outside the forested area: in CALM's Greenough, Wheatbelt and South Coast Regions. CALM's Southern Forest Region, which comprises most of the southern forests, has comparatively few rare, threatened or vulnerable taxa.

A checklist of the flora of the Porongurup Range, the largest karri outlier, contains 488 species and is considered to be 90 per cent complete (Keighery 1988b). Five species are endemic to the Range, and do not occur in the main southern forests. Several of the important understorey species of the main karri occurrence are absent from the flora of the Porongurup Range (Abbott 1982).

Ferns and Mosses

The ferns of the area are not well known (Hopper *et al.* 1992). Nineteen species are listed in Appendix 1, most of which are widespread species that occur outside the south-west. The occurrence of a few individuals of the tree fern *Sphaeropteris cooperi* in isolated localities is interesting. It is likely that this species is introduced (Christensen *et al.* 1985).

The mosses of southern Australia are described in Scott *et al.* (1976). No current list exists for the area. Two known mosses are of interest: a new subspecies of the famous Tasmanian biological rarity *Pleurophascum grandiglobum* is known from two sites on Two Peoples Bay Nature Reserve just to the east of the southern forests; and the endemic species *Rhacocarpus webbianus* is known only from Mt. Chudalup and Two Peoples Bay Nature Reserve.

Lichens and Fungi

Sammy (1985) lists 193 species of lichen for Western Australia. Inspection of his lists and locality records suggests that 79 species (41 per cent of the total number of known species in Western Australia) occur in the southern forests (see Appendix 4). This total includes 15 species which have been recorded from the Porongurup Range, an outlier of the southern forests. This very high concentration of lichens in the southern forests is also illustrated in Sammy's (1985) distribution maps.

Hilton (1982, 1988) has completed a census of all known species of larger fungi in Western Australia based on citable herbaria specimens. The list, which includes the pioneer collections of Drummond, Preiss and von Mueller, includes some 504 species, comprising 12 families of Ascomycetes and 51 families of Basidiomycetes. Inspection of the locality records provided suggests that 135 species (27 per cent) comprising seven families of Ascomycetes and 36 families of Basidiomycetes have been recorded from within the southern forests (see Appendix 5). This is a high proportion of the total number of Western Australian species.

Fauna

The disappearance of native mammals following European settlement is an Australia-wide phenomonen from which the forested areas have suffered less than the more arid areas (Burbidge and McKenzie 1989).

Christensen *et al.* (1985) recorded 27 species of native mammals (including 9 bats), 145 species of birds, 32 species of reptiles, 15 species of frogs, and 12 species of fish in the southern forests. These figures represent respectively 96, 79, 56, 75 and 100 per cent of the species in each of the groups that occur in the forested areas of the south-west.[1] In addition, 10 species of introduced mammals and five species of introduced fish occur within the area (see Appendix 3).

Only 5 per cent of the vertebrate species are restricted to the southern forests (Christensen *et al.* 1985). Most species also occur in the northern jarrah forest, and eastern wandoo woodlands.

[1] Calculated from survey data presented in Christensen *et al.* 1985.

Table 2b

SUMMARY OF PLANT TAXA WITH PRIORITY FOR CONSERVATION BY CALM MANAGEMENT REGIONS

REGION	DRF[a]	PRIORITY CODE[c] 1	2	3	4	5	TOTAL
Northern Forest	27	33	36	35	7	20	158
Southern Forest	14	13	16	11	1	7	62
Central Forest	18	24	18	19	2	20	101
Metropolitan	9	25	6	15	2	4	61
Gascoyne	5	19	15	12	0	4	55
Goldfields	6	30	19	17	2	7	81
Greenough	65	118	78	76	11	26	374
South Coast	73	128	154	89	11	29	484
Wheatbelt	81	96	82	56	13	21	351
Pilbara	3	9	9	6	0	4	31
Kimberley	3	16	24	16	0	2	61
State[b]	**247**	**456**	**390**	**252**	**53**	**106**	**1504**

a DRF: Declared Rare Flora 17 May 1991.
b Species may occur in more than one CALM Region.
c Priority codes as of 14 December 1990.

Note: for administrative purposes CALM's estate in W.A. is divided into 11 management regions (see CALM 1988).

Table 2b *(continued)*

PRIORITY FLORA LIST

For Western Australia
(19 December 1990)

Key:

[c] PRIORITY CODES

Priority 1 - Poorly known taxa.

Taxa which are known from one or a few (generally <5) populations which are under threat, either owing to small population size, or being on lands under immendiate threat, e.g. road verges, urban areas, farmland, active mineral leases, or the plants are under threat, e.g. from disease, grazing by feral animals. May include taxa with threatened populations on protected lands. Such taxa are under consideration for delaration as 'rare flora', but are in urgent need of further survey.

Priority 2 - Poorly known taxa.

Taxa which are known from one or a few (generally <5) populations, at least some of which are believed not to be under immediate threat (i.e. not currently endangered). Such taxa are under consideration for declaration as 'rare flora', but are in urgent need of further survey.

Priority 3 - Poorly known taxa.

Taxa which are known from several populations, at least some of which are believed not to be under immediate threat (i.e. not currently endangered). Such taxa are under consideration for declaration as 'rare flora' but are in need of further survey.

Priority 4 - Taxa presumed extinct.

Taxa which have not been collected, or otherwise verified, over the past 50 years despite thorough searching, or of which all known wild populations have been destroyed more recently.

Priority 5 - Rare Taxa

Taxa which are considered to have been adequately surveyed and which, while being rare (in Australia), are not currently threatened by any identifiable factors. These taxa require monitoring every 5 to 10 years.

Christensen *et al.* (1985) used multivariate analysis to examine the distribution of vertebrate fauna recorded on 25 separate biological surveys in the south-west forest areas of Western Australia. In particular, the analysis examined whether zoogeographic zones could be defined. They concluded that,

> In spite of the relationships between survey groupings and the botanical districts of Gardner and Bennetts, it would be difficult to argue the case for a distinct and separate vertebrate fauna in any part of the south-west land division.
>
> The results of the analysis of the surveys suggest a continuum of fauna distributed along a temperature/moisture gradient, which is particularly strong north/south and weaker in an east/west direction. The more xeric adapted groups, such as the reptiles, are more frequent in the warmer, drier northern forest areas. The mesic fauna, such as amphibia and fish, reach a peak of development in the cooler, moister south.

This is supported by How *et al.* (1987) who show that the coastal herpetofauna of northern Busselton-Margaret River area differs from that further to the south and east. They also demonstrate a positive correlation between the percentage of viviparous reptiles in subregional assemblages and the annual number of rainy days, perhaps suggesting that vivipary confers reproductive advantage in a cool, moist climate.

Another characteristic of the fauna is that a high percentage, especially among the higher vertebrates, are dependent on habitat elements provided by trees in a mature or derelict state, or by fallen logs. Thus, many species of mammals and birds require tree or log hollows for nesting or shelter. This is important with respect to conservation in an area where timber harvesting is a major activity (Wardell-Johnson and Christensen 1992).

Mammals

The area's fauna of 27 native mammals includes five macropods, four possums, four dasyurids, the southern brown bandicoot, the numbat, the echidna, two rodents, and nine bats. In addition, a total of 10 species of introduced mammals (including the dingo) occur in the area (Appendix 3a).

In summarizing findings of a study of the ground vertebrate fauna of coastal area between Busselton and Albany, How *et al.* (1987) suggested that mammal populations are small and often isolated, giving as examples the ringtail possum (*Pseudocheirus occidentalis*) and the quokka (*Setonix brachyurus*). This appears also to be the case with a number of other species in the area. This may not always have been the case; the introduced European red fox (*Vulpes vulpes*) has reduced populations of many species to low numbers.

Six mammal species are endemic to the south-west of Western Australia. They are the honey possum (*Tarsipes rostratus*), the quokka (*Setonix brachyurus*), the western brush wallaby (*Macropus irma*), the western ringtail possum (*Pseudocheirus peregrinus occidentalis*), the grey-bellied dunnart (*Sminthopsis griseoventer*) and the bat *Fallsistrellus mackenziei*, all of which occur within the southern forests. The honey possum is the sole member of the monotypic superfamily Tarsipedoidea. The dentition of this species is diprotodont and it would seem to be a phalangeroid, but it has no close relatives among the possums and has therefore been placed in a superfamily of its own (Strahan 1983). Honey possums occur in woodland and heathland areas but are more common in kwongan vegetation outside the southern forests.

The quokka is most common in dense karri and coastal shrubland communities (Christensen *et al.* 1985). Apart from its short tail, this species looks much like any other wallaby, but details of its dentition, skull structure, chromosomes and blood proteins segregate it from the typical wallabies in the genus *Macropus*. It is a browser and may be related to the stock from which the forest wallabies and the tree-kangaroos arose (Strahan 1983).

The western brush wallaby is one of the largest and most attractive of the wallabies. It is still common in parts of the area, although it is absent from the true karri forest and the south coastal areas, and is declining in overall numbers. It has not yet been the subject of detailed research.

The western ringtail possum, a subspecies of the common ringtail[2] (Strahan 1983) that occurs in the eastern States, has a patchy distribution within the area. It is most common in the jarrah forests of the Perup in the north-east of the area, and small colonies occur in riverine habitats in the karri forest and in peppermint woodland on the coast. This species is uncommon over much of its range.

Recent morphological studies by Kitchener *et al.* (1984) have resulted in the splitting of the former species *Sminthopsis murina* into five species. One of these, *S. griseoventer*, has its main occurrence within the southern forests, but it also occurs throughout the south-west and on the south coast.

In a revision of the Australo-Papuan *Pipistrellus* and *Falsistrellus*, Kitchener *et al.* (1986) have recognized the former south-west form of *Pipistrellus tasmaniensis* as a new species, *Falsistrellus mackenziei*. The species is endemic to the south-west forested areas and a few adjacent locations. It was one of the bats collected most frequently on surveys in the southern forests by Christensen *et al.* (1985).

Two species, the numbat and the woylie (*Bettongia penicillata*) have become 'recent endemics' owing to contraction of their range since European settlement. The numbat, the mammal emblem of the State of Western Australia, is the only living member of the Myrmecobiidae. It is one of the very few species of mammals in the world which lives solely on termites, and it is seriously endangered. It now occurs only in a few forest and woodland areas of the south-west and one of its major strongholds is the Perup Nature Reserve. It is one of the most beautiful of our native animals and one of the few diurnal species. It is the subject of intense research and protection management.

The woylie *(B. penicillata)* in the southern forests is also the subject of an intense study at the Perup Nature Reserve (Christensen 1980b). This is the major centre of occurrence of the subspecies[3] in the southern forests. The diet of this species illustrates the intricate coevolution and interdependence of indigenous species. Woylies can use hypogean fungi, mostly mychorrhizal species, as a food source. Being deficient and imbalanced as a source of amino acids, fungi are not a good source of food for mammals, so the woylie does not use the fungi directly. Bacteria in the large sacculated fore-stomach digest the fungi and these bacteria and their by-products, which constitute a more balanced diet, are digested in the posterior part of the stomach and small intestine (Kinnear *et al.* 1979).

The chuditch, also common in the Perup Nature Reserve is the subject of study at present (Serena *et al.* 1991). It is another species with a shrinking distribution which appears to have its main distribution within the forested areas.

Four of the endemic or 'recent endemics', the numbat, the woylie, the western ringtail and the chuditch, are also declared as threatened fauna i.e. are 'likely to become extinct, or are rare' (Wildlife Conservation Act 1950-1979). Another species on this list, which is also common in the Perup Nature Reserve, is the tammar wallaby (*Macropus eugenii*). These days, it persists only on islands off the shores of Western and South Australia, and in a few mainland population in south-west Western Australia. The Perup Nature Reserve has the largest mainland populations of this species.

---

[2] The western ringtail is believed by some authorities to be a separate species (Burbidge personal communication, Department of Conservation and Land Management, W.A.)

[3] Many authorities do not accept the subspecies status of this species (Burbidge personal communication).

A sixth species recently added to the threatened list is the southern brown bandicoot (*Isoodon obeselus*). This species is common in the Perup and throughout the south of the area in places where there is dense understorey, particularly along the coast. Of the aforementioned species, four are of special concern because of their comparative rarity outside the southern forests. These are the woylie, with a population estimated not to exceed 5000, the numbat, estimated total population of less than 1000 animals (Christensen *et al.* 1985)[4], and the quokka and western ringtail possum.

The Perup Nature Reserve populations of the first two species are critical to the survival of the species. The quokka populations in the southern forests, although small and isolated, are comparatively numerous and represent a large proportion of the surviving population of this species. The western ringtail possum has its major occurrences in the area from north of Busselton through to Dunsborough and at the Perup. A substantial portion of these populations, as well as other smaller occurrences, including one at Albany, occur within the southern forests (Jones and Howe[5] unpublished report). The chuditch and the western brush wallaby are all in low numbers and may be declining.

Of the remaining species of mammals, the yellow-footed antechinus or mardo (*Antechinus flavipes*) is common throughout the southern forests on sites which have remained unburnt for 10 or more years. The area represents the stronghold of the species in Western Australia; it also occurs along the eastern seaboard.

The nine species of bats that occur within the area are all common, particularly in the more open forest and woodlands.

Two species of mammal no longer occur in the area, the boodie (*Bettongia lesueur*) recorded from the Lake Muir area prior to 1911 (Kitchener *et al.* 1978), and the dalgyte (*Macrotis lagotis*), found near Bridgetown in the Blackwood River valley prior to 1933 (Western Australian Museum Nos M001399 and M001749). In both cases their habitats are now largely cleared for farming and there is little chance of finding colonies of these species in the area. The former is now considered to be extinct on the mainland; the latter survives in desert areas in northern and central Australia.

Another species which appears to be extinct in the area, but which could conceivably still occur in dense swamp vegetation within the area, is Gilbert's potoroo (*Potorous tridactylus gilberti*). John Gilbert obtained specimens from Albany in 1843, and in 1866-1869 further animals were obtained by George Masters. Gilbert's potoroo and the broad faced potoroo (*Potorous platyops*) were the subject of an intensive search in 1975 and 1976 by the Department of Fisheries and Wildlife (Kabay and Start, 1976; Start 1982). Although they did not locate any potoroos, their report concludes that the species may still inhabit the area. Subfossil remains of *P. tridactylus* are common in sand dunes on the west coast in the southern forests, but no sign of this species was found by Christensen *et al.* (1985).

There is a single record of *Phascogale calura* from Bridgetown in 1910. This species had a widespread distribution in the arid and semi-arid parts of southern and central Australia, and it is unlikely that it was ever abundant in the southern forests. It still occurs in a few areas of remnant vegetation in the Western Australian wheatbelt.

Pseudomys shortridgii, the heath rat, occurs in scrubland and heaths near Ravensthorpe and in the Fitzgerald River National Park, and has never been recorded in the southern forests in modern times.

---

[4] The populations of both of these species have increased since this estimate was made.

[5] Jones and Howe unpublished report W.A. Museum.

In summary, the southern forests have an interesting mammal fauna, in particular the north-eastern jarrah-wandoo woodlands of the Perup Nature Reserve which contain several rare and restricted species. The Perup Nature Reserve has populations of all extant mammals recorded within the area, with the exception of the honey possum and the quokka. The survival of so many of the medium sized marsupials in the Perup Nature Reserve is believed to be owing primarily to the presence of heartleaf poison *Gastrolobium bilobum*. This plant contains a poison which is lethal to introduced predators such as the European fox, but to which the native fauna have developed a high level of tolerance (Oliver *et al.* 1977). The effectiveness of fox control and fire management programs in the area since the 1970s have also contributed.

It should be noted that the Perup Nature Reserve is not typical of much of the southern forest, being jarrah and wandoo woodland in the 650 to 700 mm rainfall zone. The fauna, particularly the mammals, are more typical of the semi-arid zone. It is included in this review for completeness and geographical coverage. It is a very important proposed nature reserve for conservation of endangered medium-sized (<500 g) mammals which was left out of consideration in the chapter on reserves in the book on the northern jarrah forest (Dell *et al.* 1989).

Birds

A total of 145 species of birds have been recorded as occurring in the southern forests (Appendix 3b). This represents 79 per cent of the birds listed for the forest areas of the south-west (Christensen *et al.* 1985). If some of the waterbirds and waders (estuarine species), as well as a few introduced species, are subtracted from the total given for the south-west forests in Christensen *et al.* (1985), then the percentage of forest birds recorded in the southern forests rises to almost 90 per cent. This suggests that the southern forests are of major importance to south-west forest birds. Apart from this, the avifauna of the area has comparatively few features that distinguish it from the avifauna of the rest of the forested areas of Western Australia.

In general terms, the open woodland and low open woodland formations in the area support the highest number of bird species. The high open and open forest formations contain lesser numbers of species, but the number of individual birds is generally higher; this is especially so in the karri forest. The high open forests are also the stronghold of many of the Pssittaciformes (parrots and cockatoos).

A number of birds are endemic to the south-west, but none appear to be endemic to the southern forests. Southern forests are nonetheless considered to be important to several species.

Two species which are largely confined to the forested areas of the south-west are the endemic forest subspecies of the red-tailed black cockatoo (*Calyptorhynchus magnificus naso*) and the endemic species of the white-tailed black cockatoo (Baudin's or the long-billed white-tailed black cockatoo) (*Calyptorhynchus baudinii*), which is confined to the >750 mm rainfall area. An isolated south-west population of the widespread barking owl (*Ninox connivens*) also occurs in the southern forests.

Two other species of south-west endemic parrots, the western rosella (*Platycerus icterotis*) and the red-capped parrot (*Purpureicephalus spurius*) are also common in the area. The western rosella is particularly numerous in the high open karri forest; the red-capped parrot is most common in the open jarrah forest and in the karri forest. *Purpureicephalus* is the only endemic avian genus in the south-west. Areas of forest containing mature trees that provide hollows for nesting are particularly important for all of these parrot species.

The red-winged fairy wren (*Malurus elegans*), which is endemic to the south-west, is common throughout the southern forests wherever dense understorey exists, especially in the high open forest. The white-breasted robin (*Eopsaltria georgiana*), and the red-eared firetail (*Emblema oculata*) are also endemics to the south-west, their range extending well outside the forested area into coastal woodlands and heaths. Both of these species are nevertheless particularly common in the southern forests.

There are also a number of birds with disjunct populations occurring in the south-west forest and woodland areas as well as in similar habitat in the east of Australia, e.g. the white-naped honeyeater (*Melithreptus lunatus chrysoptera*), the spotted pardalote (*Pardalotus punctatus*), the scarlet robin (*Petroica multicolor*) and others (see Slater 1974; Blakers *et al.* 1984).

Some species which are common in the southern forests represent birds which have differentiated into eastern and western Australian forms. Thus there are eastern and western Australian forms of the ring-necked parrots, the mallee ringneck (*Barnadius barnardi*) in the east, and the Port Lincoln ringneck (south-west form), (*Barnardius zonarius semitorquatus*) in the south-west (Blakers *et al.* 1984).

Rare bird species include the ground parrot (*Pezoporus wallicus*), with the last recorded sighting in the southern forests being made at Irwin Inlet in 1952 (Serventy and Whittell 1976). This species is now rare in Western Australia. Small populations survive further to the east of the area in coastal heathlands, and the species may still survive in the area near West Cape Howe (Blakers *et al.* 1984) where there have been unconfirmed sightings in recent years. Lewin's water rail (*Rallus pectoralis clelandi*) the south-west form of the more widely distributed *R. pectoralis* formerly occurred in the area but has not been sighted in recent times.

Suitable coastal habitat exists in the southern forests for four of the State's rarest birds, the noisy scrub-bird (*Atrichornis clamosus*), the bristle-bird (*Dasyornis brachypterus*), the Rufous bristle-bird (*Dasyornis broadbenti*) and the western whipbird (*Psophodes nigrogularis*). All four species occurred within the area in historical times, but no sightings have been reported in recent years. Suitable habitat (e.g. the coastal swamps and thickets) still exists in near pristine condition in several of the coastal national parks. It is possible that one or both of the bristle-birds may still be found in the area. An active program of re-establishment of the noisy scrub-bird is in progress at the present time (Burbidge *et al.* 1986).

The noisy scrub-bird is particularly interesting. It and the closely allied Rufous scrub-bird of eastern Australia are anatomically one of the more primitive groups of the songbirds. They are the only known songbirds that lack a furculum or 'wishbone', the clavicles being reduced to mere splints, evidently a consequence of its near flightlessness (Serventy and Whittell 1976).

Reptiles

Reptiles as a group are poorly represented in the southern forests. A total of only 32 species were recorded by Christensen *et al.* (1985) (Appendix 3c). No dragon lizards (Agamidae) have been recorded in the southern forests. Geckoes (Gekkonidae), goannas (Veranidae) and legless lizards (Pygopodidae) are particularly poorly represented, the former two groups by only one species each and the last by three. One aquatic tortoise (Chelidae) occurs in the area. Nine snakes, one species of Typhlopidae, one of Boidae and seven Elapids, were recorded in the area by Christensen *et al.* (1985). Of note are two endemic and rare species, the square-nosed snake (*Rhinoplocephalus bicolor*) and the short-nosed snake (*Notechis minor*). Both species are confined to coastal woodlands and sandy areas to the south of the karri forest. The short-nosed snake, which is largely confined to the southern forests, seems to be particularly rare and is very seldom sighted. The square-nosed snake is more common but it is confined to a narrow area south of the karri forest and east along the coast to the Thomas River.

Skinks (Scincidae) are the best represented group of reptiles in the area, a total of 19 species being recorded by Christensen *et al.* (1985). A large number of these skinks are endemic to the south-west, including *Egernia luctuosa, Egernia napoleonis, Egernia pulchra pulchra, Ctenotus labillardieri, Ctenotus catenifer, Ctenotus impar, Hemiergis initialis initialis, Hemiergis peroni peroni, Lerista microtis microtis, Sphenomorphus australis, Leiolopisma trilineatum,* and *Morethia lineoocellata.* Of special interest is *Sphenomorphus australis,* an uncommon skink whose centre of distribution is within the southern forests.

Amphibians

Fifteen frogs, two species in the family Hylidae and the rest in the family Leptodactylidae and all endemic south-west species, were recorded in the area by Christensen *et al.* (1985) (Appendix 3d). To this must be added two further species of *Geocrinia,* recently discovered and named (Roberts *et al.* 1990). This total represents some 77 per cent of the total number of frogs listed for the south-west forest areas (Christensen *et al.* 1985). The southern forests are therefore the most important area in the south-west for frogs.

The genus *Geocrinia* comprises five species, including the two recently described species. The *Geocrinia rosea* complex with four allopatric species, *Geocrinia rosea, G.lutea, G. alba and G. vitellina* is entirely restricted to the area. The fifth, *G. leai,* occurs largely within the area, but its distribution extends to Mundaring in the northern jarrah forest.

The four allopatric *Geocrinias* have very limited distributions. *Geocrinia rosea* is confined to the valley systems of three rivers, the Donnelly, the Warren and the Shannon. *Geocrinia lutea* is confined to certain swamps in the karri, and tingle forests in the vicinity of Walpole. The two recently discovered species are even more restricted, the first being found in swamps to the east of the Boranup karri forest, and the other being restricted to one area of approximately 20 ha (Wardell-Johnson and Roberts 1991a, 1991b).

Fish

Twelve species of native fish were recorded for the area by Christensen *et al.* (1985) (Appendix 3e). With the exception of the mullet (*Mugil cephalus*) and the pouched lamprey (*Geotria australis*), all species are endemic to the south-west. There is also one endemic monotypic family, the Lepidogalaxiidae and two endemic genera, *Nannatherina* and *Bostockia.*

With the exception of the black-striped minnow (*Galaxiella nigrostriatus*), the Swan River goby (*Pseudogobius alorum*) and the western minnow (*Galaxias occidentalis*), the endemic fish occur largely in forest streams. Several of these, the salamander fish (*Lepidogalaxias salamandroides*), the mud minnow (*Galaxiella munda*), the night fish (*Bostockia porosa*), and Balston's pygmy perch (*Nannatherina balstoni*), have very resticted ranges.

The minnows (Galaxiidae) are well represented in the area, and of the five known south-west species, only *Galaxias truttaceous* was not collected in the area. This species is known to occur just east of the area in the vicinity of Albany (McDowall and Frankenburg 1981), and its range very likely extends into the area.

Invertebrates

Aquatic Invertebrates

The number of crustacea of the southern forests is not known. Riek (1967) has reviewed the freshwater crayfish of Western Australia and lists four species. Two species of these are endemic to the south-west and two are restricted to the southern forests.

Since that time there have been considerable advances in the knowledge of aquatic invertebrates of the area, and the following information was gained by personal communication from scientists working in the field and largely supersedes that in Reik (1967).

Austin (1986) recognizes six native species of *Cherax* in Western Australia. Five are native and endemic to the south-west: *C. crassimanus*, *C. glaber*, *C. preissii*, *C. quinquecarinatus* and two subspecies of *C.tenuimanus*. *Cherax destructor*, now widely distributed throughout the wheatbelt and probably absent from the area, was introduced from eastern Australia.

C. crassimanus, *C. glaber and C. tenuimanus tenuimanus* are restricted to the southern forests. *C. crassimanus* is distributed along the south coast from Margaret River in the west to the Hay River in the east. *C. glaber* is known only from the area between Dunsborough to just east of the Donnelly River. *Cherax tenuimanus tenuimanus* has now a very restricted distribution, being known only from Margaret River.

At least three and possibly four distinct south-west species of terrestrial crayfish of the genus *Engaeura* occur in the area. All three species are restricted to within 30 km of the coast from just east of Walpole to Dunsborough - a distribution that virtually restricts them to the southern forests.

The possible fourth species, which is electrophoretically quite distinct, is restricted to a small area in the south coast to the east of Denmark. The taxonomy of this group is still being investigated (Horwitz[6] personal communication).

There are other species of aquatic invertebrates of interest: *Daphnia occidentalis* is restricted to a very small area near Mt. Chudalup in the D'Entrecasteaux National Park. The small freshwater shrimp *Paleomonetes australis* is another endemic south-west crustacean common in fresh water streams in the southern forests (Christensen 1982). The taxonomic status of this species is uncertain as the group is in need of revision. It has been suggested that it may be congeneric with *Macrobrachium intermedium* (Boulton and Knott 1984) with which it occurs in the Swan River.

Pusey and Edwards (1990) list the macroinvertebrates recorded from karri forest streams and pools associated with the southern acid peaty flats (sedgelands). The list, which they consider to be incomplete, includes Mollusca (3 spp.), Annelida (1 sp.), Arachnida (11 spp.), Crustacea (29 spp.) and Insecta (108 spp.). Many of the species listed have not yet been properly identified.

The main point to come from their work to date is that the aquatic macroinvertebrate fauna of the southern acid peaty flats is similar to that of the jarrah forest, but there are additional species which do not occur in jarrah forest streams.

A list provided by Growns[7], which he compiled from sampling karri forest streams near Pemberton, lists Bivalva (1 sp.), Crustacea (12 spp.) and Insecta (63 spp.). This list contains only two species of Crustacea and seven species of Insecta in common with the list of Pusey and Edwards (1990) from the southern acid peaty flats. The present evidence therefore suggests that the karri forest macroinvertebrate stream fauna is different from that of the southern acid peaty flats, and from that of the northern jarrah forest.

Spiders

Main identified 18 families of spiders from collections made in the karri forests in the Pemberton area (see Curry *et al.* 1985). This includes the mygalomorph *Chenistonia villosa*, which is confined to karri forest habitat.

[6] B. Horwitz, University of Tasmania.

[7] I.O. Growns, Murdoch University, W.A.

Insects

Abbott (in press) provides a preliminary list of named insect and closely allied species definitely known to occur within the forested part of south-west Western Australia. All relevant journals and books available in institutional libraries in Perth were searched.

The list includes 1743 named species occurring in 228 families in 24 orders. The biogeographical implications of the list are of particular relevance. One quarter of all species so far recorded occur in both the northern and southern forests, one third occur only in the northern forests, and one half occur only in the southern forests. Approximately 23 per cent of insect species are known only from the karri forest. The southern forests species are listed in Appendix 3f.

There are also differences shown by various orders in the patterns of distribution of species. There is a clear trend of increasing species richness in the southern forests of the orders Odonata (dragon flies), Coleoptera (beetles), Diptera (flies) and Tricoptera (caddis flies). Other orders have more species present in the northern forests and some are distributed fairly evenly. The most speciose families show similar differences with the Lygaeidae, Carabidae, Dytiscidae, Staphylinidae, Scarabaeidae, Tenebrionidae, Cerambycidae, Chrysomelidae, Curculionidae and Geometridae having the most species recorded from the southern forests.

An interesting feature of the list is that about half of the species are known from only one locality. Abbott suggests that this may be a measure of patchiness of species distribution in south-west forests.

Abbott also presents evidence to suggest that the total number of insect species present in the forested part of the south-west Western Australia could be in the order of 15 000 to 20 000. These findings suggest that many more species remain to be found in the southern forests and the karri forest itself may be particularly rich in insect species in relation to its total areas.

Scenic Features and Wilderness Values

In addition to biological and geological attributes, the karri forest has aesthetic values which are an essential consideration in management. The southern forests are unique as the mildest, best watered corner of a large State that is noteworthy for its harsh aridity. Shaded gullies concealing fresh, clear, permanent streams abound. Cool nights entice sleep, and mornings are often overcast with a suggestion of mist, which clears by mid-morning. This contrasts with the harsh light and the sting of the sun so typical of summer elsewhere in southern Western Australia.

Many descriptions of the virgin karri forest border on the extravagant. Terms such as 'sombre magnificance' and 'cathedral-like' are appropriately used. The contrasting colours of the light, green-tiered crowns and darker green shrubs and understorey against the salmon pink and grey of the tree trunks, especially when viewed in the slanting light of early morning or late afternoon present an unforgettable sight. Walking along a track through 'old-growth' karri forest is an uplifting experience, of which few people ever weary.

Apart from the forests, the area contains other splendid landforms with landscapes of exceptional natural beauty. Several places deserve special mention. The first spans the mid to lower catchments of the Deep and Frankland Rivers across the Kent River to Mt. Lindesay. Here, craggy granitic peaks offer magnificent views over a landscape of remarkable diversity: meandering treeless flats change abruptly at their edges to woodland and low forest communities; further up-slope these in turn change to towering karri forest, through which other granitic crags protrude. Across this general pattern the valleys of large streams are discernible, with here and there a glimpse of deep permanent water glinting through massed karri crowns. Diversity of vegetation and landform, and the rugged granitic outcrops, together contribute to a sense of untouched naturalness, providing a rare sense of primaeval wilderness.

The remote and lonely coastline between Augusta and Walpole is a second area of special significance. Here, the landscape varies from long windswept beaches, towering sea cliffs (Plate 11), craggy headlands and low coastal heaths on spectacular steep dunes, to closed peppermint forest with its inviting dappled shade (see Plate 5), to forests of jarrah, marri and yate. In places, the karri forest intrudes along the major streams and in protected localities. Sometimes, as at the Walpole Inlet and Lake Maringup, it comes down to the water's edge.

On this coast there is a profound sense of wilderness with the sky, the sea, the wild surround, and the ever present brooding karri forest on the skyline. There is also a strong sense of history because this was a coast well known to the Bibbulmen Aborigines and later the early European navigators, whalers and sealers, many of whom were escaped convicts from Van Diemen's Land. Over the centuries these people trod these same hills perhaps with a different purpose, but with probably much the same sense of awe as we do today. Remarkably, the wildness of this coast (now wholly within, or proposed for inclusion in the Walpole-Nornalup and D'Entrecasteaux National Parks) remains intact.

Old-Growth Forest

In recent years the term 'old-growth' forest has been marketed to conjur up an image of untouched, primaeval stands, a place where huge old gnarled and stately trees festooned with lichen and lianes reach for the sky through a tangled undergrowth of strange and unknown species. Among the leafy canopy and in the deep litter among the rotting logs of huge, old veterans of bygone days, scamper rare and endangered species, often new to science. A conseqence of this fairytale image of the forest is that rational debate on utilization and reservation of 'old-growth' forest has become extremely difficult.

To compound the problem the term 'old-growth' forest has a different meaning in Australia from North America, where it originated. Old-growth eucalypt forest in Australia is essentially what is referred to by foresters as 'virgin' or 'mature forest'. In North America the term is increasingly being used to describe what has long been referred to by botanists and foresters as the later seral stages in forest succession. In the Pacific Northwest for example if forests of huge Douglas fir are left undisturbed for 100 to 150 years growth slows and the ancient trees decline and die; eventually Douglas fir is replaced by hemlocks, cedars and silver firs. A wildlfire will reverse the cycle. The young forest is dominated by alder which in turn is replaced by Douglas fir as the forest matures (Dayton 1990).

The biological significance of this succession is that certain species, for example, the northern spotted owl, may be dependent on the later seral stages of the Douglas fir forest. In the absence of extensive areas of these old fir forests with an abundance of dead and dying trees, the owl may disappear (Dayton 1990). Eucalypt forest in Western Australia does not progress through a series of seral stages; the understorey may do so in some cases, but the forests themselves are the 'climax' stage. There is an exception to this in Tasmania where mixed eucalypt-rain forest types revert to rain forest in the absence of fire for a period of several hundreds of years (Ashton 1981). A fire returns the forest to a eucalypt-dominant stage.

Therefore 'old-growth' eucalypt forest does not have the same biological significance in Western Australia as it may have in some other places. Old-growth forest *per se* is therefore not a critical factor in biological conservation in Western Australian eucalypt forests. This is not to say that there are no species which have requirements which may be provided only by mature or senescent trees. There are very many hole-nesting fauna species, for example, which require trees of an age that have developed suitable holes (Wardell-Johnson and Christensen 1992).

These species, however, do not depend on old-growth forest in the same way as the spotted owl. Many can inhabit heavily cut-over forest, and some even small copses of trees on farmland, as long as they have access to suitable mature trees.

A major difference between the forests of the south-west of Western Australia and those of south-eastern Australia is that unlike the south-eastern forests, they contain no gliders. Furthermore, most mammal species which live in the forest are not restricted to forests. The greater proportion of Western Australian forest mammals occur also in woodlands, heathlands, mallee and even in the deserts.

In the south-west forests, as in much of the rest of Australian eucalypt forests, it is the fire-climax or late seral stages in the undergrowth which may be biologically important. The high fire incidence which has prevailed in the south-west for many thousands of years (see Hallam 1975) has left very few species which cannot cope with a regime of frequent fires. The noisy scrub-bird, which was restricted to one small area of long unburnt south coastal scrub, has increased in numbers following a reduction in fire frequency. Some spiders of the tingle forest (Main B.Y. 1987) may also be relicts of an earlier age when fire was a less frequent natural event. It is important to ensure that representative unburnt areas are provided in the major forest biomes (see Christensen and Abbott 1989).

Biological factors aside, there is a social requirement for the reservation of old-growth forest. Most people admire natural mature forest for its beauty and interest, and it has a scarcity value in Australia as elsewhere. Public perception of old-growth forest often includes long-unburnt understorey as a part of the definition, i.e. the requirement is for virgin forest, long-unburnt with an abundance of dead material, rotting logs, fungi and mosses. In Western Australia we are fortunate that there remain many substantial areas of karri, jarrah, tingle, yate and other forests which meet these requirements (see Table 6).

Two Western Australian old-growth forests deserve particular mention: the tingle forest near Walpole and the karri forest in the Warren National Park.

There is a special feel about the tingle forest with its huge towering trees, with massive buttressed trunks and the dense tall understorey growing down to the sea. This forest is one of the most wonderful features of the southern forests.

The karri forest of the Warren National Park, which is in the heart of karri country, is some of the finest forest in existence. It was recognized as outstanding when it was first discovered, and then reserved early this century. As well as the magnificent karri trees themselves, there is the steep topography of the Warren River valley, and the dark deep river itself, often flanked by banksia and Warren River cedar *(Agonis juniperina).* The area has all one might expect of old-growth forest.

There are many other fine stretches of forest and smaller stands that have been set aside in conservation reserves in the southern forests.

In many of the forests managed by CALM, areas have been set aside as reference areas managed to exclude fire (Christensen and Abbott 1989). For such areas to remain unburnt for long periods requires detailed planning and active protection, and only a small percentage of areas can ever be set aside in this way. Numerous smaller areas exist in places where they escape fire, being protected by natural features such as rocks, lakes and sand dunes. Some plant communities, such as swampy areas, burn infrequently and seldom catch alight even when the forest around them is in flames. Unburnt fire-climax areas have special values and constant and concentrated effort is required to maintain them.

Summary

Only about 20 per cent of the southern forests have been alienated and cleared for agriculture, and nearly half the remaining forest is managed primarily for nature conservation. Few species are likely to have become extinct since European settlement, and surveys indicate that the area still supports an almost complete[8] flora and fauna. This is unusual in modern Australia. The flora is rich, with a very wide range of plant associations, and endemic species. The fauna is virtually intact with very few species having become extinct in recent times; *Potorous tridactylus gilberti* and *Rallus pectoralis clelandi* are two known examples of species that have become extinct. There are a number of rare and endangered species among the mammals, and many species with restricted distributions. (It should be borne in mind that it is normal for the majority of species in an ecosystem to be rare and localized.) The mesic element, frogs, fishes and aquatic invertebrates in particular, are of interest.

The area also has some outstanding scenic features, not least of which is the karri forest itself. There are also still areas of high wilderness values, particularly on the south coast and to the east of the Frankland River. There also exist considerable areas of forest which is still 'old-growth' or virgin forest in the sense that it has not been cut for timber (see Chapter 7).

---

[8] In the sense of what was there in 1829.

CHAPTER 3

MAJOR INFLUENCES ON THE FOREST ENVIRONMENT

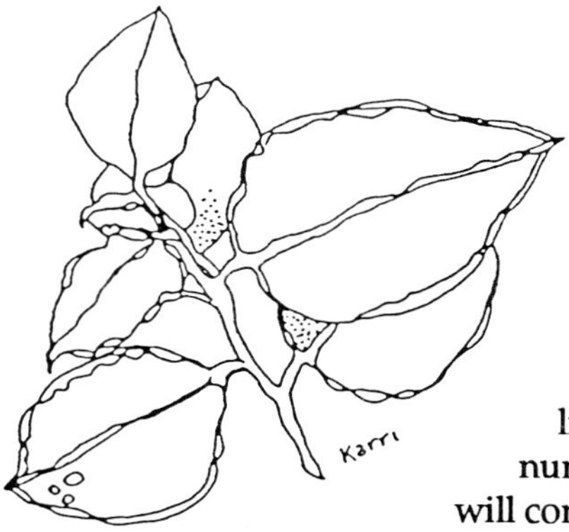

Prehistoric and more recent historic influences have all contributed towards shaping the present environment. Some of these influences, like the burning which was practised by the Aborigines, no longer occur, but the results of these practices are, and need to be, emulated if the pre-European environment is to be conserved. Other influences like the introduction of the fox and cat, plant diseases and numerous aggressive weeds are having a deleterious effect and will continue to do so for a long time to come.

Many influences result from the human use of the area. People have left evidence of their presence - historical things - some of which need to be preserved. Links with the past are an important part of the human psyche which contribute to our general well being. Past and present influences are therefore an important consideration in the conservation of the area's present environment.

Fire - An Integral Part of the Natural Ecosystem

In the introduction to his study of the biology of the woylie (*Bettongia penicillata)* and the tammar (*Macropus eugenii)* in relation to fire, Christensen (1980) states

> It is likely that fires have occurred since combustible materials first appeared on earth. Lightning, as an expression of the earth's electro-magnetic field, has been a constant source of ignition, and a major fire agent in recent times (Komarek 1967 and 1968), and there is every reason to believe that it was so in the past. American coal beds provide evidence of ancient fires during the Carboniferous and later epochs (Komarek 1972), and there is also evidence of prehistoric fires from Mesozoic coals in Australia (Gill 1973). The role of fire in ecosystems has been extensively reviewed by Kozlowski and Ahlgren (1974), and there can be little doubt it is a major force of selective adaptation in many ecosystems.

There is a close association between fire, eucalypts and their associated xeromorphic understoreys (Mount 1964; Mutch 1970; Webb 1970). Xeromorphic taxa are considered to have first evolved in the early Tertiary, radiating during the pluvial period when laterite was formed, and forming mosaics with rain forest which then occupied sites of higher fertility (Beadle 1981).

As the Australian continent drifted northwards during the Tertiary and the arid zone developed, the wet scleromorphic communities were replaced by communities adapted to drier conditions. From the mid-Miocene, the flora adapted and migrated in response to the general increase in dryness. During the Pliocene it seems probable that the widespread rain forest of the early Tertiary was replaced by vegetation dominated by such genera as *Eucalyptus* and *Casuarina* (Kemp 1981).

Recher and Christensen (1981) consider that fire has been an important factor in the evolution of the Australian biota since the early Tertiary. In recent years the study of fire in Australian ecosystems has received particular attention, see reviews by Gill (1975), Recher and Christensen (1981) and Gill *et al.* (1981) and for Western Australia, see reviews by Ford (1985) and Christensen and Abbott (1989). Fire in the karri forest is discussed by Christensen and Annels (1985).

The importance of fire in the continuation of ecological processes, and in maintaining xeromorphic ecosystems like some of the forest, woodland and heathland communities of the southern forests, cannot be in doubt. Examples exist of intricate interrelationships between plants, animals and fire. Christensen (1980) described the 'fire dependence' in the tammar wallaby and showed how a particular fire regime is essential to the survival of this endangered species in the Perup Nature Reserve. He also outlined the intricate relationships between the woylie, fire, certain species of hypogean mycorrhizal fungi and scarab beetles.

There exist many more examples of adaptations that enable plants and animals of the area to cope with a very wide variety of fire regimes. Fire in the past may well have been a major factor in determining the ecosystems which we see today with their associated species of plants and animals. We may ask, for example, why is it that the cool, temperate rain forest element, so prevalent in south-eastern Australia, is missing from the southern forests of Western Australia. Where, for example, are genera such as the southern beech, *Nothofagus*, once widespread over the continent during cooler wetter periods. *Nothofagus* is still a component in eastern forests of *Eucalyptus regnans*, *E. delegatensis* and *E. nitens*, trees which have many characteristics in common with karri.

One explanation why cool, temperate rain forest elements are absent from the southern forests, is fire frequency. Cool, temperate rain forest is known to replace tall, open eucalypt forests in the prolonged absence (hundreds of years) of fire (see Ashton 1981). Fire in the karri forest is a much more frequent event (see Underwood 1978; and Christensen and Annels 1985). This idea is supported by the general absence of a suite of k-selected dependent resident mammals from the karri, which can be interpreted as a consequence of the relatively high fire frequency in this forest (see Christensen *et al.* 1985). (Species which display k-selective traits, viz., longevity, late sexual maturity and low fecundity, tend to favour habitats which are relatively stable and not subject to frequent disturbance by fire, for example.)

In eastern Australia, high, open forests similar to the karri, but with a lower fire frequency (Ashton 1981), have living in them a range of dependent resident mammals including the koala (*Phascolarctos cinereus*), pygmy glider (*Acrobates pygmaeus*), Leadbeaters possum (*Gymnobelideus leadbeateri*), sugar glider (*Petaurus breviceps*), yellow-bellied glider (*P. australis*), greater glider, (*Petauroides volans*), and mountain possum (*Trichosurus caninus*) (Tyndale-Biscoe and Calaby 1975). These species tend to display typical k-selective traits such as longevity, late sexual maturity and low fecundity and be adapted to a comparatively narrow range of habitats.

The mammal fauna of the southern forests contains no dependent residents, and the species which do occur there generally display r-selective traits, namely short life span, early sexual maturity and high fecundity. These are traits which are associated with species living in frequently disturbed environments, for example, those frequently disturbed by fire (see Christensen *et al.* 1985).

On the other hand, Main B.Y. (1987) suggests that the persistence of certain invertebrates, e.g. relict spiders (*Chasmocephalon* and others) in the karri-tingle forests, suggests that fire in this habitat is inimical. Main B.Y. (1987) postulates that some spiders of the tingle forest are relicts of a period prior to the advent of fire as a component of the ecosystem and that they persist only where fire is a rare phenomenon, or at least where unburnt refugia remain within burnt-over areas. Main[9] (personal communication) has subsequently found other species from other spider groups in the Walpole-Nornalup area which gives further support to her hypothesis.

Burning by Aborigines for some 40 000 years (Hallam 1975) has undoubtedly eliminated many fire sensitive species. These few species have successfully managed to escape this fate.

[9] Dr. B.Y. Main, University of W.A.

Fire has obviously been a factor in shaping the present communities, and it will need to be taken into account if we wish to conserve the present ecosystem and natural processes.

In the 40 000 years of their occupation of the areas before the 19th Century, Aborigines imposed fire regimes on south-west forest ecosystems. They used fire extensively, to clear the undergrowth, for hunting purposes, to create habitat for suitable food species, and for many other reasons (Hallam 1975). Aborigines probably did not use the thick karri forest areas to any great extent, but they are known to have inhabited inland woodlands during winter and spring, and the coastal areas, particularly during the warmer and drier months of summer. Their burning activities certainly affected the forest areas (Hallam 1975). Aboriginal fire practices would have helped shape the present environment, and this needs to be taken into account, particularly in managing coastal reserves where their influence appears to have been the greatest.

Humans

The Early Inhabitants

Aborigines arrived on the Australian continent more than 40 000 years ago (see White and O'Connell 1982). Their lifestyles and activities differed from place to place depending on the culture of the people concerned and the regional variation in the environment and resources. Information on the early inhabitants of the southern forests comes from a number of archaeological sites that date from the Pleistocene to the Holocene, a period spanning from 30 000 to 5 000 years ago. In the south-west only very few of the known archeological sites date back to the Pleistocene. Three of these, Devil's Lair, Arumvale and Quinninup Brook, are in the southern forests. They are all in the same locality on the Leeuwin-Naturaliste Ridge; the two former on coastal limestone to the west of the Boranup karri forest; the last north of the karri on coastal heath with stunted marri. There is another possible late Pleistocene site near Dunsborough, also on the Leeuwin-Naturaliste Ridge, but further to the north (Ferguson 1981).

Devil's Lair, the oldest of these sites, has produced evidence of intermittent human occupation of the locality from 33 000 to 12 000 years ago (Dortch 1979a). Archaeological information from these sites has provided insight into the subsistence patterns and technological processes of environmental adaptations of late Pleistocene - early Holocene human inhabitants of the south-west region. The inhabitants used a variety of resources, often widely spaced, which they visited along a series of interconnecting pathways. These paths often followed streams and were kept open by burning. Camping places were at sources of permanent water where the pathways came together (Hallam 1975).

When Europeans first arrived, the Leeuwin-Naturaliste Ridge was within the tribal area of the Wardandi (Tindale 1974). There are almost no direct records of their traditional way of life. The area was more sparsely occupied than the Swan coastal plain and Hallam (1972) believes this to have been a consequence of the high rainfall which supports thick forests of jarrah and karri over much of the area inland from the coast.

Ferguson (1981) indicates how information on the habits and lifestyles of Aborigines at the time of settlement, together with archaeological data, can be used to reconstruct the past. Recent research suggests that the southern Australian Pleistocene climate was cooler and more arid than at present (see Galloway and Kemp 1981). Evidence from studies of late Pleistocene vertebrate fauna excavated from Devil's Lair and other limestone caves in the area reveals a more arid-adapted fauna from this period, which suggests that areas where there is now forest might have been more sparsely vegetated at this time (Balme *et al.* 1978).

A more open environment would have enabled the human inhabitants to move more freely and probably supported larger populations of fauna. The richest vertebrate remains in Devil's Lair excavations roughly coincides with the glacial maximum 19 000 to 12 000 years ago (Balme *et al.* 1978). At that time sea levels were 150 m lower, and there was a vast coastal plain below the present archaeological sites, an extensive area of which would have been available to the mammal fauna of the time (see Ferguson 1981).

Ferguson suggests that the grindstone assemblage in the Quinninup Brook excavations reinforce Dortch's (1974) suggestion that the south-west Early Phase tool assemblage indicates methods of exploitation which are similar to those of Aborigines found in the area in historical times. Thus they may have used a similar network of paths to the Aborigines of historical times to exploit the area, and the archaeological sites may have been a part of this network. Glover (1975) suggests that the ancient coastal plain was part of this network. The presence of fossiliferous chert at Quinninup Brook, postulated to have come from quarries on the continental shelf, supports this view. The Pleistocene sites on the Leeuwin-Naturaliste Ridge would seem to have been part of a larger exploitation network that was changed as a result of marine transgressions.

The shrinking Aboriginal exploitation area is suggested to have been coupled with a deterioration in the hospitality of the part that remained. Thus there was an increase in wet conditions during the terminal Pleistocene-early Holocene which peaked 7 000 to 5 000 years ago (Rognon and Williams 1977). Pollen studies (Churchill 1968) show an increase in the distribution of karri forest at this time. This coincides with the local extinction of a number of non-forest species of fauna in the early to mid-Holocene (Balme *et al.* 1978; Merrilees 1979; Porter 1979).

The excavation of these sites also provides an insight into the culture of these early people of the southern forests. Thus there is evidence of a flaked stone industry which included quartz, chert, granitic-gneiss and silcrete. These early people used grinding stones, and a variety of fairly sophisticated stone and bone artifacts, including bone beads (see Dortch 1979a, 1979b, 1980). They also used kangaroo and wallaby incisors for tools or ornaments (Balme 1979).

Similar artefacts are known also from further south in the southern forests near Northcliffe (Dortch and Gardner 1976). They state that

> Aboriginal stone workers used local chert, silcrete, quartz and other stone. They quarried silcrete extensively at an outcrop near Northcliffe from before 6 780 years BP until at least 3 000 years BP. Stone artefact assemblages contain diverse retouched tools made on flakes and blades, notably geometric microliths.

More recently, Aboriginal sites have been discovered on the lake bed of Lake Jasper. Investigations are continuing, but numerous stone artefacts, including microliths, have been found so far (Western Australian Museum Information Sheet).

There is some dispute over the extent of the use of shellfish as food by Aborigines because of the apparent dearth of coastal and other shell middens. However, Dortch *et al.* (1984) present evidence of marine mollusc used by Aborigines in the south-west. Most of the sites they discuss occur close to rocky shores in the area between Malimup, south of Northcliffe, and Albany, and contain gastropods characteristic of such habitats.

The south-west, including the karri forest, is noted for its conspicuous lack of Aboriginal art (Davidson 1952; Serventy 1952; Crawford 1972; Merrilees *et al.* 1973; Hallam 1981). Only three examples are described from the southern forests, a few engraved or scratched limestone plaques from Devil's Lair (Dortch 1976), some animal track engravings on the Scott River near Augusta (Clarke 1983) and some hand stencils in Old Kurdardup cave near Augusta (Morse 1984).

The latter are the only known examples of painted rock art in a south-west Western Australian limestone cave.

The first European visitors to the south-west coast of Western Australia noted stone structures which they concluded were used by Aborigines for catching fish. Today the remains of these structures are known from several places within the southern forests. At Wilson Inlet near Denmark on the south coast, there is a site consisting of several ill-defined enclosures, and at Broke Inlet there are at least eight separate enclosures that occur along less than one kilometre of the shoreline of the estuary. These structures and their possible use are discussed by Dix and Meagher (1976). In 1988-1989 Dortch[10] (personal communication) identified a complex of stone structures covering several hundred square metres at the east end of Wilson Inlet. These stone structures, interpreted by him as fish traps, are normally submerged, but were exposed during the severe drought of 1988 when estuary levels were exceptionally low.

At the time of European settlement, the Aborigines also used fish weirs which were brushwood barriers placed in stream channels. Post-settlement accounts contain descriptions of the construction and use of such traps on the Serpentine River near Perth (Paterson 1896; Hammond 1933). The remains of a possible fish trap in the southern forests is described from Blackwater Creek near Northcliffe, by Dortch and Gardner (1976).

Little evidence has been found of Aboriginal occupation of the karri forest itself. Use of the dense forest may have been minimal, as a richer and more easily obtained food source could be had among the more easily penetrated coastal fringe (see also Hallam 1975). Apart from food sources, the coastal strip provided suitable tool-making material, the basalts and volcanic glasses at Black Point, the silcretes near Northcliffe, and granulites at Malimup. The latter location provided an important bi-faced industry as well as a rich main food source (Gardner[11] personal communication)

The coastal areas of the southern forests, particularly limestone areas where there are caves, have provided most of the evidence of early Aboriginal life in the area. It seems likely that there will be more discoveries in the future that will provide further information on these early inhabitants of the area.

Recent Inhabitants

Portions of the Western Australian coastline were explored and charted by a succession of Dutch, French and British navigators during the seventeenth and eighteenth centuries. It was not until the establishment of British settlements at King George Sound in the 1820s, and later on the Swan River, and at Augusta, that mainland exploration began.

The earliest recorded explorers of the area appear to have started at Augusta. A small party of men led by Thomas Turner attempted to trace the Blackwood River to its source. As far as is known they reached no further than the present site of Nannup before turning back. During the 1840s, further exploration was carried out along the Blackwood River, and the area's potential for agriculture began to emerge. This was followed shortly by the arrival of the first convicts in the 1850s, and, as the colony started to grow, settlers spread inland to the forest regions from the coastal settlements (Christensen *et al.* 1981).

At first the settlers grazed their sheep and cattle in the forest; later they selectively cleared the land to grow pasture, crops and orchards. The more open woodland was favoured at first for pasturing stock, later clearing followed along the more fertile valley bottoms. It was soon

---

[10] C.E. Dorch, W.A. Museum.

[11] G. Gardner, Northcliffe.

discovered that blackbutt soils were among the most fertile, so these were among the first forests to be cleared. Large areas of prime karri forest were ringbarked and cleared as a result of the soldier settlement schemes following World War I. Mistakenly, it was thought that the huge karri indicated the presence of the most fertile soils.

The existence of the extensive forests with their timber potential was not realized until the 1850s and commercial utilization of the karri did not begin until the 1880s when M.C. Davies established his first sawmill at Karridale in 1882. Two years later the Millar Brothers established a mill at Torbay, west of Albany. Milling of the area's forests has continued and much of the remaining forest area has today been cut-over and regenerated at least once. Early utilization was selective; the best (tallest and biggest) trees were used first, and the intensity of cutting varied from one place to the next depending on wood quality and quantity, distance from the mill as from mill to port.

Like the Aborigines, the later inhabitants have used resources selectively so that the impact on the environment is different in different landforms and plant formations. There have been low losses of species in forests in comparison with non-forested landforms. Western Australian forests still retain a very high proportion of their original species of animals and plants.

Perhaps the greatest impact of humans has been the indirect effects of the introduction of pest and weed species which compete with, and on occasions, eliminate native species. Feral animals such as the horse, pig, goat, rabbit, fox, cat, rats, house mouse, and the kookaburra are all present in forests and impacting on forest ecosystems. Introduced grasses, particularly from Africa and the Mediterranean, have invaded much of the sandy coastal areas, and are contributing to changes in the fire regime (Christensen and Burrows 1985). Wattles from the eastern States, watsonia from South Africa, blackberry and others are having an effect on the native plants and animals.

Then there are the less conspicuous effects of introductions; fish, like the redfin perch and the mosquito fish, which may be changing aquatic environments.

Dieback disease carried by introduced fungal root pathogens, *Phytophthora* spp., especially *P. cinnamomi*, is perhaps the most serious conservation problem in the southern forests as it is in many south-western forests, parks and reserves (Dell *et al.* 1989; Shearer and Tippett 1989; CALM 1990a; Shearer and Batini 1990).

The disease is known to occur in many places throughout the southern forests. Most infections are along roads and tracks but in southern areas, on the southern flats in particular, infection has spread over considerable areas by water flow.

The fungi infect a wide range of susceptible and resistant host plants; however, they only flourish in susceptible plant tissue. Susceptible species occur primarily among the families Proteaceae, Epacridaceae, Xanthorrhoeaceae, Myrtaceae, Dilleniaceae and Papilionaceae. Infections may slow the rate of growth of less susceptible plants in other families.

Susceptible plants may be killed, and in communities with a high proportion of susceptible species, it is not uncommon to find considerable areas, often several hectares in extent, where a high proportion of the plants are dead. With time the plant composition of such areas changes, resistant monocotyledonous species often becoming more dominant. Such changes, which are likely to be permanent, may affect animal communities. For example, a range of nectar feeders rely heavily on many of the susceptible species, e.g. the honey possum.

Many plant communities which have susceptible species, and occur on moist or water gaining sites, and on soils which impede drainage, are under threat. Such threatened communities are most frequent in the southern parts of the forest and along the coast. In the Walpole-Nornalup National Park where more detailed studies have been done, many plant communities have been found to have a high proportion of susceptible species, and to occur in sites with a high risk of infection and spread. Several species of priority flora (see Table 2b) were shown to be susceptible and at risk (CALM 1990a).

Lastly, the increasing salinity caused through land clearing, of some of the streams and the rising saline ground water, particularly in the drier eastern sector of the area, is a major change caused by human activity (Anon. 1989).

Summary

Past influences have helped to shape the present environment. Some of these past influences, for example, natural and perhaps also aboriginal fire regimes, may need to be reintroduced into certain ecosystems in order to conserve these ecosystems and their natural processes. Other changes, because of their selective nature, e.g. clearing for agriculture, have limited the options for conservation by reducing or even eliminating certain communities, species and habitats.

Potentially perhaps the most damaging are the influences which are still actively at work, perhaps only just starting to have an impact. In this catergory are the introduced pests, diseases and weeds. They are the new colonizers. The fox gives a clue to their potential effect. Three species of mammal are extinct in the southern forests and three are declining in numbers (see Table 5), a situation almost certainly owing to the predation of the fox. Other introduced species move more slowly: *Phytophthora cinnamomi* and some of the weed species, for example, are only just be beginning to exhibit their full and serious potential.

These are the factors which have been, and continue to be, major influences on nature conservation in the area. The selection of a reserve system and the ongoing management of the reserves is to a very great extent restricted by these past and present influences.

Evidence of past human influence in the area is still to be demonstrated. Where this is of significance and where practical, it should be included in the reserve system for its historical and other values.

CHAPTER 4

CONSERVATION SIGNIFICANCE OF THE SOUTHERN FORESTS

The Evolution and Development of the Australian and South-West Biota

Significant geological events

What is the conservation significance of the southern forests flora and fauna? While many conservation biologists promote conservation of all species, it is nevertheless useful to distinguish those species that seem to be more important and have more significance in the context of land management for conservation. To appreciate the true value of anything, one needs to examine it in relation to other things; in particular related or similar things. Biology is no different. The conservation significance of species is best measured by comparison with other species. How common are they, how widespread, are they of new or ancient lineage? Species which are uncommon, with restricted distribution, and which originate from the far distant past, have a high conservation significance, for example the small fish *Lepidogalaxias salamandroides* and the noisy scrub-bird. Species which are common and widespread, and which have evolved in more recent geological time, for example, the lesser long-eared bat (*Nyctophilus geoffroi*) and the galah (*Cacatua roseicephala*) have less conservation significance.

Judgements on the conservation value and significance of species such as the examples given above are not made on their present distribution and abundance alone. It is necessary to have an understanding of their phylogenetic relationships as well as the relationship of the species with others on a continental and global scale. The southern forests biota is therefore examined in the wider context of an Australian and world biota, as well as its evolutionary and geological history. This is achieved using several temporal and spatial scales from the distant past through to the present, and from the continental down to the regional scale.

The world's major landmasses, including Australia, were not always distributed across the globe in the way they are today. All the continents were once joined together in a single huge land mass, Pangaea, which, over a long period of geological time, broke up into segments and separated, finally to become the present continents (see Keast 1981b).

This supercontinent divided into the northern Laurasia (Eurasia and North America) and southern Gondwana (Antarctica, the southern continents, the Indian subcontinent, Australia and New Zealand), separated by the Tethys Sea in equatorial regions, except at the western end of the Mediterranean.

The breakup of Gondwana was a Cretaceous (156 to 78 million years ago) event, starting with the separation of North America from Antarctica approximately 156 million years ago. Australia maintained connection with Africa and New Zealand through South America and Antarctica until the late Cretaceous 78 million years ago, and South America through Antarctica until 49 million years ago when Australia finally separated from Antarctica. Direct migration from Antarctica via Tasmania and the south Tasman rise may still have been possible until about 3 million years ago (Kennett 1978; Raven 1979).

Time of separation of North and South America is of great biogeographic interest with respect to Australia because marsupials are considered to have originated in North America, diversified during the Cretaceous and migrated to Australia via South America and Antarctica.

Australia continued its northward movement through 15° latitude for about 30 million years, experiencing marked changes in climate, until it collided with the Asian tectonic plate in the region of the Sunda Arc. This took place in the mid to late Miocene, about 15 million years ago. New Guinea formed the northern extension of the Australian plate during its northward journey, although for most of the time it was totally submerged (Dow 1977; Axelrod and Raven 1982). Land had commenced to appear in New Guinea at the end of the Eocene (Dow 1977) but it remained insular until the collision of Australia with the Sunda Arc in the Miocene approximately 15 million years ago.

When Australia separated from Antarctica and drifted northwards 50 million years ago, it carried with it elements of both the Pangaean and the Gondwanan biota. As the continent drifted northwards evolutionary and adaptive changes in Australian plants and animals must have accelerated as warmer latitudes were approached and the continent became drier.

The significance of these events in the interpretation of the present day flora and fauna is summarized by Keast (1981a)

> Australia had a rich and diversified biota in the Palaeozoic and Mesozoic, with significant segments shared much of the time either with Pangaea, or the rest of Gondwana. There was, of course, progressive extinction of the floras and faunas of these periods on all continents. Ceratodontid lungfishes, however, extend back to the Devonian in Australia, at least nine orders of insects to the Permian, some of the conifer genera to the Jurassic, and *Nothofagus* to the Cretaceous. Such forms have continued to occur ever since. Interpretations based on contempory plants and animals must, hence, be made in the expectation that not only the nucleus of the present biota but most major orders may have been present in the Mesozoic.

Main A.R. (1987) more specifically points to the drying of Australia caused by the development of global ice sheets after the Oligocene as the event that was the direct cause of selective pressures which ultimately led to the adaptive radiation of the Australian fauna. He further highlights the Cretaceous fragmentation of the present continent into a number of islands separated by epicontinental seas as an event which produced disjunct distribution of fauna. This allowed the south-west fauna (Main A.R. 1987) and flora (see Nelson 1981) to develop in isolation from the rest of Australia.

Development of the Australian and the South-West Flora

Evidence suggests that angiosperms first appeared on earth not much more than 12 million years ago during the Cretaceous (Hickley and Doyle 1977). Diversification apparently occurred rapidly, and a significant number of taxa had arisen before the breakup of Gondwana began about 120 million years ago (Mueller 1970; Schuster 1972, 1976; Raven and Axelrod 1974; Johnson and Briggs 1975). Thus the flora of Australia contains all the major plant families of the world with the exception of the two largely American families, Cactaceae and Bromeliaceae (Beadle 1981).

Following the breakup of Gondwana, the flora continued to evolve in Australia. Palaeobotanical evidence suggests that closed forests akin to the present rain forests of the tropics were spread over much of the continent by the Eocene, 50 million years ago. By this time the Australian flora had lost its connections with South America and Africa. Migration between Australia and South America via Antartica was, however, still possible at least for cold-tolerant species until about 40 million years ago (Raven and Axelrod 1974).

Following separation with Antartica when Australia was rafting northwards towards warmer latitudes, there was a general global cooling and, with the formation of the Antarctic ice sheets, an increase in temperature gradients from the south pole to the equator. The general cooling was accompanied by an increase in aridity (Kemp 1981) and the rain forests contracted to the east coast where more hospitable moister environments persisted (Walker and Singh 1981; Lange 1982). Tiny pockets of rain forest persist in favourable areas of the Northern Territory, and in the Kimberley area in Western Australia's north.

As a result of the changing climate, the vegetation changed, and major evolutionary radiation took place. In the Proteaceae, sclerophyll types related to *Banksia* and *Dryandra* appeared (see Beadle 1981; Johnson and Briggs 1981). The Myrtaceous genera, the *Eucalyptus, Callistemons* and the *Leptospermums* appeared at this time (Lange 1982). Through the Pliocene, closed myrtaceous forests, unlike the modern open eucalypt forests, developed. Open eucalypt forests did not become widespread until the Pleistocene (Martin 1981).

Hooker (1859) first proposed grouping of the Australian flora based on biogeography. Nelson (1981) and Beadle (1981) proposed modified schemes to allow for new interpretations based on an Australasian biogeography. Both these authors proposed subdivisions based primarily on the time when the Australasian plate severed land connections with Antartica, approximately 50 million years ago. Both authors recognize a 'Gondwanic element': plants that were present *circa* 50 million years ago, and an 'intrusive' element. Beadle (1981) separates the 'Gondwanic element' into an ancient Gondwanan flora, the xeromorphic flora, and the arid zone flora, both of the latter derived from Gondwanan lineages. Nelson's (1981) suggested subdivisions of the 'Gondwanic element' are essentially similar;

- a relict sub-element of taxa generally confined to moist, cool semi-tropical habitats (e.g. closed forests) which have not greatly diversified into genera or endemic species since the breakup of Gondwana. This element is more common in tropical and semi-tropical habitats like the northern Queensland rain forests (Anon. 1986).

- An 'autochthonous sub-element', Australian largely xeromorphic genera which evolved from taxa which responded to changed environmental conditions since Australian isolation.

The xeromorphic segment of the flora includes the sclerophylls such as the dry-fruited Myrtaceae e.g. *Eucalyptus, Leptospermum, Calytrix,* most Acacias, Proteaceae, Epacridaceae, Papilionaceae (especially Podalyrieae), some Euphorbiaceae and many smaller taxa, as well as the soft-leaved plants with xeromorphic features, especially Rutaceae, Goodeniaceae, Stylidaceae and others. They share a preference for soils of low fertility and constitute the *Eucalyptus* forests, woodlands, mallees and heaths found mainly in the wetter parts of the continent (Beadle 1981).

The development of the autochthonous flora with its many endemic genera and species appears to have occurred during the time of rafting northwards of the Australasian plate during the Tertiary. The process is considered by Beadle (1981) to have started at the margins of rain forests along gradients of fertility from high to low. The widespread formation of laterites, which bound up soil phosphates during a wet period possibly in the Miocene, and the later reweathering during a subsequent pluvial period, were possibly the periods of widest distribution of xeromorphic taxa. Xeromorphic communities may have formed a mosaic with the rain forests of the time, being concentrated, as they are today, in the south of the continent.

The changes that occurred during the Tertiary effectively concentrated and isolated the major sections of the autochthonous xeromorphic flora in the south-west. Beadle (1981) states,

> The contractions of the wet xeromorphic assemblages was initiated by the development of the arid zone which effectively cut the xeromorphic assemblage into two main segments - one in south-west of Western Australia, where the flora is almost entirely xeromorphic, the other on the Triassic sandstones centered around Sydney, where small patches of rainforest still occur.

This separation of the south-west flora had already commenced with the marine incursion of the Nullarbor in the late Eocene (approximately 50 million years ago), which occurred prior to the commencement of the arid phase. As the sea retreated, the Nullabor karst, composed largely of limestones and calcarian sandstones, presented an impenetrable barrier to the largely calcifuge southern flora (Nelson 1981).

There are further reasons for the uniqueness of the south-west flora. Six areas of the world have a mediterranean climate: the Mediterranean region itself, parts of California and Chile, the Cape region of South Africa, south-west Western Australia and parts of South Australia. In only three of these, the Cape region and the south-west and parts of South Australia is there a well developed assemblage of scleromorphs peculiar to low-nutrient soils. In the other regions the flora is less diverse and distinctive and was derived by modification of subtropical woodland floras rather than from floras pre-adapted to a mediterranean climate, as is the case of the Cape and south-west and South Australian floras (Johnson and Briggs 1981). In these cases the flora was further modified by additional speciation and some physiological adaptation, superimposed upon their existing summer-growth pattern (Specht 1969; Johnson and Briggs 1975). With very few exceptions they did not spread into the adjacent moderate to high nutrient areas of mediterranean climate, which in Australia carry open forest or woodland and have much poorer woody flora of mostly trees and scrubs (Johnson and Briggs 1981). This agrees with Beadle (1954, 1966, 1981) who considers low fertility to be the primary factor in the development of new species and genera in the Australian autochthonous flora. The ultimate in low fertility is the habitat occupied by heaths especially those in the south-west. The new genera which evolved there are nearly all endemic (Beadle 1981).

Hopper (1979) lists another factor suggested by Raven (1973) as being responsible for the evolution of a rich endemic flora, the occurrence of numbers of species-specific hymenopteran pollinators, as being worthy of future investigations. Many species-specific thynnid wasps are known to pollinate pseudocopulatory orchids in the south-west (Stoutamire 1983).

The Development of the Australian and South-West Fauna

Among the fauna, elements that relate to the various stages of the separation of the continents may also be recognized. In a general statement on zoogeography of Australian insects, MacKerras (1970) suggests that they represent four biogeographical elements:

- archaic, composed of both widespread and relict groups;
- a southern element, which can be regarded as 'antarctic' or Gondwanan, comprising a very large part of the Australian invertebrate fauna;
- an old northern element containing many primitive forms representing affinities with India, Africa and Madagascar; and
- a younger northern element having mostly entered Australia via New Guinea.

In an account of the biogeography of terrestrial invertebrates, Main (1981) summarized the disparate material which is available using MacKerras' faunal elements. Prior to the breakup of Pangaea in a landscape of *Nothofagus* and *Podocarpus* and associated vegetation, she proposes an invertebrate fauna dominated by litter, humus and soil dwelling, soft bodied groups with wide geographic range.

The fauna would have been representative, but without the diversity of Pangaean elements. Relicts persist in the Onychophora (e.g. *Peripatoides gilesii*, which survives in the Darling Range in the south-west of Western Australia and in South Australia) among the spiders (e.g. *Hickmania*, which occurs in Tasmania), and among the Oligochaeta. The syncarid crustacea have been here since at least the Mesozoic (Williams 1981), all families of land snails were differentiated before the breakup of Gondwana (Bishop 1981), and the major orders of insects were also established by then. Main A.R. (1987) suggests that termites, e.g. *Mastotermes* and *Stolotermes*, are possibly survivors of an ancient Pangean element.

With the separation of Pangaea into Laurasia and Gondwana, a more distinctive southern fauna appeared. Numerous South American, New Zealand and a few old African species testify to this (some forms are common to all the southern continents and New Zealand, e.g. the trap door spider family Migidae). Generic relicts remain which have affinities with New Zealand as far back as the Cretaceous, e.g. Gradungulidae (eastern Australia and Tasmania only) and others among the spiders and some earthworms. At this time much of the fauna derived from southern connections underwent new radiations within Australia. The radiation of various groups with south-west/eastern affinities probably occurred at this time, e.g. the mygalomorph spiders Teyl/Naurea and Neohomogona/Homogona (B.Y. Main personal communicaiton) and various scarab beetles.

Numerous disjunct communities suggest that progressive drying of the continent has restricted humid-dwelling forms to the south-west and south-east 'corners' and eastern rain forest region of the continent.

Main (1981) concludes that the oldest elements in the fauna are confined to Tasmania, south-eastern Australia and the south-west of Western Australia, but notes (personal communication) that certain eastern seaboard rain forests also are geographic regions which retain some biological affinity with South America and New Zealand. The youngest elements of the fauna are restricted to north-east Queensland.

Like the plants, most faunal groups with southern representatives have many generic and specific endemics, but few have endemic families. Among the arachnids, for example, only the spider family Hickmaniidae (Tasmania only) is endemic.

The vertebrate representatives of the old Pangaean fauna include the lungfishes and osteoglossomorph fishes. An ancient endemic fish family of uncertain relationships, which is probably Pangaean and which is confined to the south-west of Western Australia, is the monospecific family Lepidogalaxiidae (Rosen 1974). Main A.R. (1987) suggests that the Diplodactyline geckos and the Pygopodidae (legless lizards) are derived from Pangean elements isolated by the fracturing of Gondwana.

The southern or Gondwanan element includes those groups exclusive in their contemporary and/or fossil distribution to the southern hemisphere; a large proportion of these include families and genera with distinct distribution between Australia and South America. Among these are the leptodactylid and hylid frogs (seven genera of the former and one of the latter groups are represented in the southern forests), chelid freshwater tortoises (represented in the south-west by the endemic *Pseudemydura umbrina*, a species now on the endangered list) and *Chelodina oblonga* (also endemic to the south-west), and marsupials. Australia and South America also share ratite birds and galaxiid fishes, Geotriidae (*Geotria australis*), the Lamprey, as well as certain Platyhelminthes (flat worms), *Onycophora* (Peripatus) and Parastacidae, Isopoda and Amphipoda (Crustaceae) with New Zealand. These groups are well represented in the southern forests (see Main A.R. 1987; and Main personal communication).

There are also very many groups of insects, especially aquatic insects, which are represented in the southern forests, and which have Gondwanan affinities: *Odonata* (dragon flies) (9 spp.) (Theischinger and Watson 1984; Watson and O'Farrell 1985), Plecoptera (stone flies) (2 spp.), Megaloptera (alder flies) (2 spp.), Trichoptera (caddis flies) (3 spp.) and Lepidoptera (moths and butterflies) (1 sp.) (Common 1990).

Among both reptiles and birds, the northern element derived from Asia is dominant. With few exceptions, all avian, lizard and snake families and genera clearly have northern relationships. The Pygopodidae, which has its greatest diversity in the south-west, is an endemic family related to the geckoes.

The Conservation Significance of the Biota

The South-West Flora

The south-west has long been recognized as a biogeographic subregion. There have been at least 20 biogeographic schemes put forward for Australia. They differ in detail but generally agree on a northern, an interior and a southern subregion. There is also general agreement on a south-west province or subdivision of the southern subregions (see Keast 1981).

The south-west province is particularly noteworthy for its autochthonous xeromorphic flora, an Australian element of ancient lineage which was mentioned earlier. Typically, Australian genera have by far the greatest diversity here (Gardner 1942; Burbidge 1960). Twenty-five per cent of the plant genera are restricted to the south-west, and the three major families, Proteaceae, Myrtaceae and Leguminosae, have a significant concentration of species in the south-west (Burbidge 1960). Species endemism is extremely high and has been calculated by various authors at between 67 and 87 per cent (see Nelson 1981).

Among the endemic genera of the south-west are *Dryandra* (Proteacae) with 60 species, *Calothamnus* (Myrtaceae) (25 spp.), *Gastrolobium* and *Jacksonia* (Leguminosae) (45 spp. and 30 spp. respectively) (Nelson 1981). Gastrolobium is especially interesting because plants of this genus contain the vertebrate toxin sodium fluoroacetate. This has resulted in the evolution of high levels of tolerance to sodium fluoroacetate among a range of native herbivores and omnivores (Oliver *et al.* 1977).

The great diversity of many of the more widespread genera in the south-west corner is even more impressive. Among the Proteaceae, *Dryandra* is restricted to the south-west; a total of 110 of the 250 spp. of *Grevillea* are similarly restricted, as are 90 of the 140 spp. of *Hakea*, and 60 of the 70 spp. of *Persoonia*, 30 of the 32 spp. *of Adenanthos*, 25 of the 35 spp. of *Isopogon*, and 30 of the 50 spp. of *Conospermum*. In the Myrtaceae, 100 of the 150 spp. of *Melaleuca*, 50 of the 70 spp. of *Baeckea*, 30 of the 35 spp. of *Darwinia* and 30 of the 45 spp. of *Calytrix*, are so restricted.

The very large number of species that occur in the south-west and the high level of endemism at the species level, are comparable with the Cape Province of South Africa (Good 1974). High rates of endemism in floras are generally regarded as indicative of long periods of isolation. Island floras generally exhibit high species endemism, e.g. Hawaii and Madagascar (see Carlquist 1974; Good 1974.). The Cape Province and the south-west of Western Australia are generally accepted as 'ecological islands' (Nelson 1981).

Hopper (1979) notes that since the pioneering work of Diels (1906), biogeographers have recognized that the autochthonous, xeromorphic, south-west flora of ancient origin is richest in the coastal and inland sandheaths of the transitional rainfall zone, and relatively species poor in the high rainfall forests and arid zone.

Species richness in most woody genera of the dicotyledons (Hopper 1979; Keighery 1984a) and in many of the monocotyledons (Keighery 1984b), is greatest in the heathlands of the south-west. The major centres of diversity occur in the heaths and scrubland of the Stirling and Barren Ranges on the south coast (Speck 1958; Marchant 1973; Nelson 1975; Keighery[12] personal communication) and the heathlands near Mt. Lesueur to the north of Perth (Griffin and Hopkins

[12] G.J. Keighery, Department of Conservation and Land Management, W.A.

1985; Anon. 1990). These areas also acted as refugia during earlier incursions of the sea in these regions (see Carlquist 1974; Main 1981).

In the south-west there are many species with very restricted geographic ranges. Many of these occur only on the peaks of the large granitic monadnocks or mountain ranges. Some of these, the Barren and the Stirling Ranges, are regarded as centres of diversity and refugia (Speck 1958; Marchant 1973; Nelson 1975).

Restriction of species to very small geographic ranges are recorded for a range of south-west genera, including *Stylidium* and *Darwinia* (in the latter five species are restricted to separate peaks in the Stirling Ranges) (Carlquist 1974), *Eucalyptus, Banksia* and *Acacia* (Seddon 1972; Nelson 1978), *Adenanthos* (Nelson 1981) and *Daviesia* (Griffin and Hopkins 1985).

The Southern Forests Flora

The area contains the most extensive area of high open forest formation in Western Australia. All the tingle forest occurs within the area, as well the large majority of the karri forest. Extensive areas of high open forest of jarrah and marri also occur as well as some yarri. In addition there are other plant formations, e.g. red flowering gum woodlands which are unique to the area.

No area with a floral richness comparable to the heathlands of the transitional rainfall zone exists within the southern forests. However, Keighery (1984b) shows that several groups of monocotyledons, i.e. Restionaceae, Iridaceae, Juncaginaceae, Juncaceae, Xyridaceae, Hypoxidaceae, Centrolepidaceae and Poaceae are species rich in the high rainfall zone of the south-west.

The high rainfall zone, on a relative area basis, also has disproportionately more monotypic relict genera (Hopper 1979). Hopper examines the reasons for this pattern, and concludes that it is the result of a series of geological circumstances. These include the preservation of some early Tertiary landscapes, the existence of marine, edaphic or climatic barriers, the extensive formation of lateritic soils that, with weathering, gave rise to nutrient deficient sands, and erosional dynamism and recurrent climatic stress in the transitional rainfall zone during the Tertiary and Quaternary. Habitat and climatic stability were greater in the high rainfall zone during these periods, favouring the persistence of a more conservative relict flora. This would apply in particular to the southern forests which have the highest rainfall in the south-west.

Although no areas with species richness comparable to the heathlands occur within the southern forests, perhaps because most of the area is occupied by forest rather than heathland, there are centres of considerable floral diversity within the area. The first of these occurs in the south-eastern section of the southern forests, i.e. the woodlands and heathland areas to the east of Mt. Lindesay (Christensen 1980). This area is close to the junction of the three botanical districts of Gardner and Bennetts (1956), the Warren, Darling and Stirling botanical districts, and contains elements of the flora of each of these. One of the few remaining extensive areas of Albany blackbutt woodlands occurs in this area, and there is greater species diversity among Banksias and eucalypts than anywhere else in the south-west forest areas.

The other areas of special botanical interest are the Whicher Range and the Donnybrook Sunklands, areas of jarrah forest, woodland and heathland in the north-western part of the area, where several plants with restricted distribution occur. *Lambertia rariflora* is one restricted species of particular scientific importance because of its primitive floral characteristics which represent definite intermediate changes in the evolutionary pathway of the proteaceous flower (Melville 1973, 1975). Other restricted species include *Pultenea skinneri, Adenanthos detmoldii* and *Franklandia triaristata*, as well as species restricted to the area such as the *Grevillea brachystylis* group and an unnamed species of pink-flowered *Darwinia*. Other species with their main occurrence in this area are *Dasypogon hookeri, Eucalyptus haematoxylon* and the spectacular *Crowea angustifolia.* Another plant of botanical interest is the tiny *Dampiera linearis*, recognized as a single taxonomic species, but including at least two biological species, the primitive diploids which occur only in the Whicher Range and near Albany, and the derived polyploids which are more widespread (Bousfield and James 1976).

Also of botanical interest are the southern 'flats', a term used for the seasonally wet and inundated treeless sedgelands and heaths which occur in association with the southern forests. Plants such as the endemic *Reedia spathacea* occur within a relatively restricted area, and it is also a major habitat of the Albany pitcher plant (*Cephalotus follicularis*). Several of the groups of monocotyledons, which reach their highest specific diversity in the high rainfall areas are common on the flats. This habitat is conceivably most like the original, low-fertility, xeromorphic heath land that occurred in association with the rain forest in Tertiary times, and from which the present rich autochthonous flora is thought to have originated. Some of the surrounding open jarrah woodland and sandy soils contain many undescribed species of orchids (Wardell-Johnson[13] personal communication).

More recently Wardell-Johnson (personal communication) has shown the Walpole-Nornalup National Park, for its size, to be one of the richest areas for plants in Western Australia. With 640 species the Park contains 43 per cent of the known plant species in the Warren Botanical Subdistrict including 17 species endemic to that Subdistrict (CALM 1990a).

Granitic monadnocks protrude like islands throughout the southern forests. Indeed they may have been islands during earlier times of higher sea level, and may have functioned as refugia like the Stirling and Barren Ranges (see Carlquist 1974; Main 1981). These are particularly numerous in the south and have a quite distinctive flora. Species such as *Acacia triptycha, A. sulcata* and *Hemigenia* spp. are common plants restricted to this habitat, *Chamelaucium 'forrestii'* is also restricted to monadnocks. Other species of interest in this habitat nearer to the south coast include the Walpole wax (an undescribed *Chamelaucium)*, Anthrocercis sp. (Plate 19), *Banksia seminuda* ssp. *remanons* and *Banksia verticillata*.

Hopper *et al.* (1992), commenting on the Warren Botanical Subdistrict, state that

> Most of the known endemics are confined to swamplands, coastal heaths and granitic outcrops with relatively few found in forests. The majority are found in the wettest country between Denmark and Northcliffe, but some are concentrated on the Leeuwin-Naturaliste Ridge (nine taxa), and two are known endemics of the Scott Coastal Plain.

Other features of scientific interest in the south-west flora, including the flora of the southern forests, include convergent evolution in pollination in species of Proteaceae by birds (see Faegri 1965; Nelson 1975, 1981) and mammals (Rourke and Wiens 1977) between the Cape Province in South Africa and the south-west since they separated in Gondwanan times 78 million years ago. The genera involved include *Adenanthos, Grevillea, Banksia* and *Dryandra* (Proteaceae), which are pollinated by birds, and *Dryandra, Banksia, Grevillea* and *Conospermum* (Proteaceae), as well as *Brachysema* (Leguminosae) and *Darwinia* (Myrtaceae) pollinated by mammals. The honey possum, whose diet is nectar and pollen (Wooller *et al.* 1984), is common in *Banksia* and *Casuarina* woodlands and certain heathlands in the southern forests (Christensen *et al.* 1985). The pseudocopulatory orchids, some of which occur in the southern forests, are also of great scientific interest because they are pollinated by species-specific thynnid wasps (Stoutamire 1983).

The array of fire adaptations which are characteristic of mediterranean floras are also well represented in the area, and have been studied by Christensen and others (see Carlquist 1974; Christensen and Kimber 1975; Christensen 1980b; Christensen *et al.* 1981; Recher and Christensen 1981). Carlquist (1974), for example, notes that *Xanthorrhoea* spp., some *Drosera* spp., and several

[13] G. Wardell-Johnson, Department of Conservation and Land Management, W.A.

terrestrial orchids, flower mainly following fire. Christensen (1980b) has demonstrated the fire dependence of the rare and restricted tammar (Plate 12). This wallaby occurs in the Perup Nature Reserve in thickets of *Gastrolobium bilobum*, an obligate seed regenerating species which germinates to form thickets only following intense fires lit under dry conditions. The thickets provide protection for the tammar from predation, especially by eagles.

The topics of parallel evolution and evolution in a fire-prone environment are of great scientific interest and the opportunity for their study is an important facet of the southern forests.

The South-West and the Southern Forests Fauna

The south-west and the southern forests fauna is not as rich in endemics as the flora. The mobility of animals means that they usually occur over a wider area than plants. For those reasons the conservation significance of the southern forests fauna is best discussed in the wider context of the south-west fauna.

The mammal fauna, like that of the rest of the continent, is a blend of old southern monotremes and marsupials, and late-arriving bats and rodents from areas north of Australia. Among the former, the monotypic marsupial family Tarsipedidae and the macropod genus *Setonix* are endemic and represented in the southern forests by *Tarsipes rostratus* (honey possum) and *Setonix brachyurus* (quokka) respectively. *Tarsipes* is common in *Banksia* and *Allocasuarina* woodlands on the south and west coast, and *Setonix* occurs in dense streamside vegetation in a variety of vegetation formations including the karri forest.

The western brush wallaby, also endemic to the south-west, is still relatively common in the jarrah forest in the northern portions of the area, but its numbers are declining. A large proportion of the remaining colonies of the south-west endemic western ringtail possum (Plate 13) occur within the southern forests at the Perup Nature Reserve, in tuart in W.A. peppermint woodland south of Busselton and at Albany. The grey-bellied dunnart (*Sminthopsis griseoventer*), another endemic, is also common throughout the southern forests, particularly in heathland and woodland communities. Among the bats, *Fallistrellus mackenziei* occurs throughout the area and it is one of the most common bats in the southern forests (see Christensen *et al.* 1985).

Of more concern from a conservation viewpoint than the two endemics just mentioned, are the species listed as rare and in need of protection - in particular, the two 'recent endemics' (species which were formerly more widespread but now are confined to the south-west forests) the numbat (Plate 14) and the woylie. Both species are common in the Perup Nature Reserve in the north-eastern sector of the southern forests. Both species have been the subject of intensive studies (see Sampson 1971; Christensen 1980b; Christensen *et al.* 1984; Friend 1990) and their biology is well known. The Perup Nature Reserve is considered to be essential to the survival of the woylie and the numbat.

Another species which is now almost confined to the south-west forests is the chuditch which has been studied by Serena *et al.* (1991). Again this species is common in the Perup Nature Reserve.

With the exception of the ubiquitous emu, there are no Ratites, the only group of avifauna considered to be of old southern origin. The only endemics, the red-capped parrot (*Purpureicephalus spurius)*, the sole member of its genus, and Baudin's black-cockatoo (*Calyptorhyncus baudinii)*, are common throughout the area. There is also the endemic subspecies of the red-tailed black cockatoo (*Calyptorhyncus magnificus naso*), which is fairly common in the area. An isolated south-western population of the barking owl (*Ninox connivens*) also occurs in the southern forests.

The reptiles, like most of the birds, are considered to be of more recent northern origin. The only family considered to be of southern origin is the Pygopodidae which has its centre of diversity to the north of Perth and is represented in the southern forests by only three widespread and common species. The only other known old endemics are two tortoise genera, *Pseudemydura* with the sole surviving member the endangered western swamp tortoise (*P. umbrina*) (Burbidge 1981), and the oblong tortoise (*Chelodina oblonga*). The former occurs only in a very restricted area north of Perth. The latter which is common throughout the area is the sole member of one of the three species groups within *Chelodina* (Burbidge *et al.* 1974)

Studies by Christensen *et al.* (1985) and How *et al.* (1987) indicate that reptile species richness is low in the southern coastal areas. Two small elapid snakes, the short-nosed snake (Plate 15) and the square-nosed snake are largely confined to southern coastal and subcoastal woodlands and heathland (Christensen *et al.* 1985). Both are uncommon and are on the endangered list. The southern forests contain at least 12 skinks which are endemic to the south-west. Of particular interest is *Sphenomophus australis*, a small skink which is uncommon and confined to the southern forests (Christensen *et al.* 1985).

The amphibia are more interesting, all 15 frogs in the area belong to two old southern groups, the Leptodactylidae and the Hylidae. The genus *Geocrinia* (Leptodactylidae) (Plate 16) is of great scientific interest because it provides an example of speciation imposed by ecological barriers. This topic has been the subject of scientific debate since White (1978) first challenged the traditional theories of speciation put forward by Main *et al.* (1958), Martin (1972) and later Littlejohn (1981). More recently, Barendse (1984) has supplied allozyme data supporting endemic speciation in the south-west. The four species comprising the *Geocrinia* complex are all spring-summer breeders and have eggs that undergo direct development (Main 1957, 1965; Roberts *et al.* 1990). Direct development is a character not found in other *Geocrinia* species, nor in related genera such as *Crinia* (sensu Heyer *et al.* 1982), implying a close relationship between the four species (Wardell-Johnson and Roberts 1991a). Direct development is more typical of arid-adapted frogs, and its occurrence in frogs living in the moist habitats of the southern forests is of scientific interest.

The two newly described *Geocrinias, G. alba* and *G. vittelina*, both restricted to the southern forests (Wardell-Johnson and Roberts 1991b) are declared as threatened species.

The freshwater fish of south-west Western Australia are considered a distinct element of Australian fish fauna with a high degree of endemism (Whiteley 1947; Allen 1982; Merrick and Schmidt 1984). At least nine of the 17 species occur in the southern forests and almost all of these are endemic to the south-west.

Of particular interest is the small salamander fish, which belongs to the archaic monotypic family Lepidogalaxiidae (Plate 17), apparently related to the pikes (Eoscidae) of the northern hemisphere (Rosen 1974). Rosen believes that this relationship dates back to the breakup of Pangaea. If it is not an eoscid, *Lepidogalaxias* may have distinct salmoniform relationships (McDowall 1981). It seems likely that this little fish, which is common in the peaty waters of the southern flats of the southern forests (Christensen 1982) is destined to play a key role in zoogeography. In addition, there is interest in the species in its own right as a primitive form of fish.

The galaxiids, an old Gondwanan family, are represented in the area by several endemic species. Two of these *Galaxiella munda* and *G. nigrostriatus*, have their main distribution in small streams and freshwater pools in the area. Two other species of small fish, *Bostockia porosa* and *Nannatherina balstoni*, belong to genera with their centre of distribution in the southern forests.

The Porongurup Range, one of the karri outliers, which was an island during the Tertiary, preserves relicts of that time, e.g. the Mygalomorph spider *Neohomogona bolganupensis* (Main 1985). Main considers that the coastal belt of the southern forests, in particular Torbay and the West Cape Howe area and the karri-tingle forests of Walpole-Nornalup, represents a 'reservoir' that harbours several eastern Australian species and genera that no longer occur in the intervening region, e.g. *Laperousa quindecimpunctatum*, a Tasmanian species, *Austrardaea, Orsolobids* and *Hadrotarsids* and others (B.Y. Main 1981, 1987). Some of these are considered relicts of a westward migration in the Pliocene-Pleistocene. The primitive *Tasmanoonops mainae* as well as some other archaeids may be much older relicts. Main (1981, 1987) considers that West Cape Howe was an island during the Miocene and that it, along with the Torndirrup area, has conserved other very ancient invertebrate relicts. *Australobus torbay*, endemic to Torbay Head has only recently been described in the family Orsolobidae, a Gondwanan family (Forster and Platnick 1985).

The genus *Tasmanoonops* probably occurs as a relict in many wet areas of the karri forest; *T. australis* is known only from near Pemberton.

With reference to the spiders, Main, B.Y. (1987) states

> the forested south-west corner and southern coastal area of the State harbour taxa that are closely related to eastern Australian rainforest forms (Orsolobidae, Archaeidae, Anapidae, Symphytoguathoidea, and others), particularly of south-east Queensland and Tasmania, as well New Zealand, Madagascar and Chile. Many genera are clearly Gondwanan relicts. Most of these taxa are small, less than 5mm and are confined to microhabitats in litter, moss humus, under bark and tree butts and in hollows and overhangs of rotten logs. Although some of these taxa have representatives living in epiphytes or tree moss in rainforest in Queensland, in southwest Western Australia they are terrestrial (even the web-weavers), a concomitant of summer drought. Few south-west Australian relict species have been named, and several families have not been formally recorded or have been cited only in ecological papers, e.g. Archaeidae, Mimetidae and Toxopidae.

An example of a Gondwanan spider with African affinities is shown in Plate 18.

Four species, *Chasmocephalon* sp., *Dardarnus* sp., *Baolebulb* sp. and *Migas* sp., have so far been recorded only in the Walpole-Nornalup National Park (Main personal communication).

Main, B.Y. (1987) argues that some of these taxa predate the occurrence of fire as a frequent natural phenomenon in the south-west, and are now at risk from the prevalence of fire. She argues that the tingle forest (*Eucalyptus guilfoylei and E. jacksonii*) at Nornalup and Walpole is probably the successor of the now locally extinct *Nothofagus* and that as such, the tingle forest and its associated fauna is unlikely to be fire adapted to the extent of some other associations such as karri or jarrah forests. Associated with the tingle are relictual spider taxa inhabiting leaf mould humus and tree butt hollows, while others are found in microhabitats in coastal heath where summer fog is prevalent.

The cave spiders of the south-west are also of particular interest since, unlike their eastern States counterparts, they can seldom be shown to be derived directly from surface fauna. Main, (1981) considers many of them to be relicts of a wetter early Tertiary climate, e.g. the ray spiders Theridiosomatidae, found only in wet habitats if outside caves in south-west Western Australia, and similarly the Mimetidae (pirate spiders) which are only known from deep moist litter and damp microhabitats apart from caves. There are many caves in the southern forests, particularly in the Cape Leeuwin-Cape Naturaliste area. Other caves may be discovered in limestone areas on the coast further to the south which could harbour interesting fauna.

The Scarabaeidae (Coleoptera) have been relatively well studied in Australia and several old Mesozoic elements such as the Geotrupidae and the Scarabaeinae have undergone radiation in Western Australia during the Cretaceous. A high percentage of the species in these two groups are found in coastal habitats. These groups are interesting as it is considered that they have evolved cool temperature adaptions. For example, many Geotrupidae, particularly the genus *Bolborbachium*, which is well represented in the south-west corner, are active during the coolest wettest periods in July and August when temperatures may be 4° to 5°C (Howden 1981). The early radiation in relation to emergent land areas in the south-west in these groups is similar to the pattern suggested for some spiders by Main (1981) and supports her contention of the scientific importance of the coastal areas, especially elevated areas such as the numerous large granitic monadnocks.

Among the Crustacea, the Parastacidae are considered Gondwanan in origin (Williams 1981). Knowledge is incomplete, but it appears that a very high proportion of crustaceans are south-west endemics with distributions centred in the southern forests. On present knowledge, some species appear to have very limited natural distributions.

The Crustaceae and other aquatic invertebrate fauna of the southern forests need further study since they are not well known and new endemics continue to be found. Examples are the subspecies *Cherax tenuimanus teniumanus* (Austin 1982), and the electrophoretically distinct species of *Engaerva* (Howitz personal communication). Even more recently a possible new species of *Perthidia* (Amphipoda) was collected by Ivor Growns from a stream in the karri forest north of Pemberton. On the same trip he also collected an undescribed genus of mayfly previously collected by Arthur Neboiss in 1978, from Carey Brook (Davis[14] personal communication)

There are further undescribed species in Growns collections from Carey Brook. Three genera of Leptophlebidae (Ephemeroptera), two species of Tasyderidae (Diptera), (a largely tropical family previously known only from one species in the south-west) and a species of Pelecorhynchidae (Diptera). The latter is a family previously recorded only from southern Chile and eastern Australia (apparently a Gondwanan relict). Lastly, a species of Ditlectrona, Hydropsychidae (Tricoptera) a genus not previously recorded in Western Australia, was also collected.

Gondwanan relicts are still being identified in the southern forests among the other invertebrate groups. A small endangered prosobranch snail (belonging to a new genus) *Austroassiminea letha* (family Assimineidae) has been described from a few isolated freshwater seepage areas near Augusta and Margaret River (Solem *et al.* 1982). The Assimineids are common in South-East Asia, Indonesia, New Guinea and the Pacific Islands, but this is the sole Australian representative of that family. The absence of any member of the family in northern Australia suggests that *A.letha* is a Gondwanan relict. The possession of anatomical features which combine aspects of both of the recognized subfamilies of the Assimineidae suggests that it may be a phylogenetic relict as well as geographic relict.

Main A.R. and Main, B.Y. (personal communication) have compiled a list of aquatic invertebrates including insects (a few are terrestrial) with Gondwanan affinities (Table 3).

---

[14] Dr. Jenny Davis, Murdoch University, W.A.

Table 3

INVERTEBRATES WITH GONDWANAN AFFINITIES WHICH OCCUR IN THE SOUTHERN FORESTS

Aquatic invertebrates other than insects
(After A.R. and B.Y Main 1991 personal communication)
An asterisk indicates those species that occur outside the southern forests.

Taxon	Affinities	Status within the southern forests
PLATYHELMINTHES		
(flat worms)		
Temnocephalildae	Australia,	
Temnoceophala sp.	New Zealand,	Widespread parasites on fresh
	and South America	water crayfish.
CRUSTACEA		
Parastacidae	Australia,	
Cherax tenuimanus	New Zealand,	Widely distributed permanent
(marron)	and South America	streams.
*C. quiquicarinatus**		Widespread permanent streams
(gilgie)		to semi-permanent swamps.
*C. preissii**		Widespread, headwaters of rivers.
(koonac)		
*C. crassimanus**		South coast to Hay River, not
		in headwaters of rivers.
C. plebejus		Widespread, swamps
Engaewa subcoerulea		Inlet River, 24 km north-west of Walpole
(burrowing crayfish)		
ISOPODA		
Phraeotoicoidea	Australia,	
S.Ind.	New Zealand,	
*Amphisopus lintoni**	and South Africa	Coastal seeps, peaty swamps,
(shrimps)		headwater streams along south coast
Paramphisopus		
*palutris**		Temporary streams in karri forest
(shrimps)		
AMPHIPODA		
Neoniphargidae	Australia and	
Neoniphargus	Madagascar	Coastal temporary swamps
(shrimps)		
CLADOCERA	Gondwanan	
Daphnia	(Benzie 1987)	6 km south of Northcliffe and
occidentalis		11 km east of Mt. Chudelup
(water flea)		
MOLLUSCA		
Pelecypoda		
Mutelidae	Australia and	
*Westralunio carteri**	South America	Bow, Warren and Frankland Rivers

Table 3 *(continued)*

Insects both Aquatic and Terrestrial

Taxon	Affinities	Status within the southern forests
ODONATA		
(dragon flies)		
Megapodagrionidae	Australia, South America, South Africa and Madagascar.	
*Agriolestes minimus**		widespread, swamps and streams
*A. pusillus**		widespread, swamps and streams
*A. pusillisimus**		widespread, swamps and streams
GOMPHIDAE		
Gomphiniae	Gondwanan affinities (Watson and O'Farrell 1985)	
*Armagomphus armiger**		permanent rapid streams
*Austrogomphus collaris**		permanent rapid streams
*A. lateralis**		permanent rapid streams
CORDULIDAE		
Gomphomacromiinae	Gondwanan affinities	
Synyhemistinae	(Theischinger and Watson 1984)	
Hesperocordula berthoudi		Permanent rapid streams
Lathrocordula metallica		Permanent rapid streams
ynthemis cyanotincta		Permanent rapid streams
PLECOPTERA		
(stone flies)		
Griptopterigidae	Australia,	
Leptoperia australica	New Zealand,	Carey Brook,
Newmanoperia exigua	and South America	Shannon River
MEGALOPTERA		
(alder flies)	Australia,	Beedelup Falls, Warren River and
Corydalidae	New Zealand,	east of Northcliffe
Archicauliodes	and South America,	
*A. ceryulus**		
TRICHOPTERA		
(caddis flies)		
Hydrobiosidae	Australia, Neotropical	
Apsilochorema urdalum		Carey Brook, Pemberton, Beedelup, Walpole
Taschorema pallescens		Pemberton, Carey Brook, Canterbury

Table 3 *(continued)*

Taxon	Affinities	Status within the southern forests
Philorheithridae	Australia, Neotropics	Walpole
Kosrheithrus boorarus		Canterbury Brook 16 km east of Northcliffe
LEPIDOPTERA (moths and butterflies)		
Castnioidea	Australia and South America	
Castniidae	Gondwanan (Common 1990)	
Synemon spp.		throughout, in *Lepidosperma* and other sedge swamps.
Terrestrial Invertebrates		
MOLLUSCA		
Bulimulidae (land snails)	Australia and South America	
*Bothriembryon melo**		widespread, coastal, calcareous sandy heaths.
*B. kingii**		coastal sands
B. jacksonii		Deep to Frankland Rivers, rare
B. frankii		Walpole, karri and tingle,rare
B. revectus		Deep and Bow Rivers, rare
ONYCOPHERA (Peripatus)		Pemberton 8 km west (Parmelia)
LUMBRICINA (earthworms)		Shannon (Dog Pool), Chudelup
Megascolex swarbricki		Nornalup
M. syndetoporus		Pemberton

Summary

The Australian biota comprises elements which originate from the times of separation of the continent from the world's various land masses. The earliest and most ancient forms dating back to the first breakup of the continents are fewest in number but they are often of greatest scientific and conservation significance.

The northward movement and drying of the Australian landmass led to the adaptive radiation of the Australian flora and fauna. It initiated the development of the autochthonous flora with its many endemic genera and species. This flora reached its greatest development in the south-west, which has remained separated to varying degrees from the rest of the continent for a very long time. The very large number of species and high endemism of the south-west sclerophyllous flora is comparable with only one other region in the world, the Cape Province of South Africa. There are very many cases of parallel evolution, a phenomenon of scientific significance, between the south-west flora and that of the Cape, particularly with respect to bird and mammal pollination and fire adaptation. A portion of this autochthonous flora occurs in the southern forests primarily in woodlands and non-forested areas.

The special 'wet adapted' xeromorphic flora which is centred in the southern forests, has special conservation significance. It is postulated here that this flora may represent the closest surviving link with the original Tertiary tropical xeromorphic flora from which the autochthonous flora developed. This is the flora of the extensive southern flats which occur in association with the southernmost karri forest. It is the landform which also contains the greatest diversity of monocotyledonous plants in the south-west.

The south-west also has a very high proportion of species with a very restricted range. These include many relictual forms which are often concentrated on elevated areas which were once islands in an inland sea. Such landforms are represented in the southern forests by the many granitic monadnocks to the south of the main karri forest. Some of these are known to contain relictual forms.

The fauna also has a range of south-west endemics, a proportion of which are represented in, though not confined to, the area. Some are 'recent endemics' like the numbat. Most of the interest in the fauna, however, is centred around the relictual forms, particularly among the lower orders, which survive in the southern forests from earlier wetter periods. Relicts are known from among the frogs, the fish and a wide range of invertebrate phyla. As yet, the invertebrate fauna is incompletely known, but many of the few known forms are of considerable scientific significance.

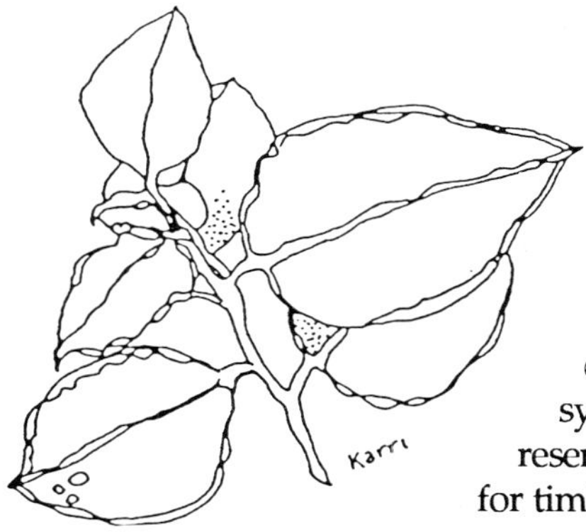

Chapter 5

CONSERVATION BY RESERVATION - SELECTION AND ESTABLISHMENT OF RESERVES AND PARKS

The Historical Background

There is a comprehensive reserve system in place in the southern forests, only a very small portion of which was originally set aside for that purpose. The bulk of the reserve system owes its existence to forestry, many of the areas now reserved for conservation having originally been set aside primarily for timber production purposes.

The existence and economic value of the jarrah and karri forests of Western Australia was recognized soon after the first European settlement at Albany in 1826. Reports drawing attention to the unbounded forests to the north and west of the settlement influenced the British Government's decision to establish a new permanent colony on the Swan River in 1829.

Commercial utilization of karri began some 50 years later. M. C. Davies erected his first sawmill at Karridale in 1882. The Millar Brothers did the same at Torbay west of Albany two years later, and in due course progressed further westward into the Denmark area. Proximity to the ports of Albany, Hamelin Bay and Augusta, were the reason for milling commencing at the extremities of the karri occurrence. Cutting in the central core did not commence for a further 30 years until extension of the railway line south from Bunbury reached Jardanup (now Jardee, near Manjimup). The provision of railway access enabled sawmills at Jardanup (Jardee, Millar Bros), Deanmill (State sawmills) and Pemberton (State sawmills) to begin operation between 1913 and 1914 immediately prior to World War I. In the meantime, exhaustion of the locally accessible resources saw closure of the mills at Karridale and Denmark. Mills at Deanmill, Jardee and Pemberton have continued in operation uninterrupted to the present day (Anon. 1969).

To accommodate a massive demand for timber in the period after World War II, new large mills were started at the Donnelly River, Quininup, Northcliffe and Shannon River in 1951. All cut karri. Of these, only Northcliffe remains in operation today, cutting jarrah, other mills currently cutting karri are located at Pemberton, Denmark, Greenbushes, Jardee and Busselton. None of the mills mentioned above uses karri exclusively, now or in the past.

Large log size and difficult access required logging operations in virgin karri to be on a large scale. Before the advent of forest management, cutting was purely exploitative and was seen as a convenient prelude to agriculture, upon which permanent settlement would depend. The sawmilling companies ran things as they saw fit. Government control was absent or minimal, and waste of the timber resource was prodigious. Despite emphatic, authoritative and commissioned advice to the contrary, and mounting unease at what was happening, the notion of forest conservation, both in a productive and preservation sense, was thrust into the background. To the early colonists, timber supplies were regarded as inexhaustible and a frontier attitude prevailed.

A major change in direction occurred during the period 1916 to 1930, beginning with the appointment of Charles E. Lane Poole as Conservator of Forests in 1916. The drafting and passing of the Forests Act in 1918 and the establishment of the Forests Department in 1919, no mean achievement by the Conservator and his supporters in the agriculture-oriented context of the times, gave Lane Poole the ammunition he needed to fight for the creation of a permanent secure forest estate, and to overcome the entrenched interests of the timber industry. Thus began an epic three-way struggle between Lane Poole, a strong and articulate advocate for forestry, Sir James

Mitchell, the then Premier and staunch proponent for agriculture, and the timber industry, which for long had had its own way. Lane Poole's resignation in 1922 was perhaps an inevitable result of his outspoken style and fierce concern for the forest. However, the battle once joined was carried on in a just as able, but perhaps more tenacious, if less heroic, style by his successor Stephen L. Kessell.

Notwithstanding the existence of the Forests Act, the declaration of a worthwhile forest estate was delayed by the continuing agriculture-forestry conflict. A change of Government in 1924, produced a cabinet more sympathetic to forestry, though its Minister for Lands was an agriculture proponent in the Mitchell mould.

Dedication of the forest estate began, and by 1930 State forests totalled approximately 1.2 million ha. The birth of the forest estate was the result of a vision shared by a few farsighted people. It was a long time evolving, often appearing as though it would never come to fruition.

Further additions to State forest were made in the period 1930 to 1954, mainly in 1938 just prior to World War II. In 1953, forest conservation received another significant boost by the formation of a Land Utilization Committee, which was required by Government to handle all matters affecting the reservation of land for timber, soil conservation and water reserves, and the release of land for agriculture and other purposes. Membership consisted of the Surveyor General (Chairman), the Director of Agriculture, the Conservator of Forests, the Principal Assistant Hydraulic Engineer of the Public Works Department, and the Commissioner for Soil Conservation. Rational land use, at least in the lower south-west of the State, benefited from its deliberations during its six-year existence (1954 to 1959). During this time a further 205 408 ha were added to State forests. In the ensuing nine years (1960 to 1969) its replacement, the Crown Lands Tribunal, added another 100 859 ha to State forests, approximately 25 per cent of the 400 000 ha requested by the Forests Department. In 1969 State forests peaked at 1.8 million ha. The area has since been reduced owing to conversion of some State forests to national parks and nature reserves.

By the 1970s most of the karri forest in public ownership was classified as State forests. Only a small percentage was in conservation or nature reserves, notably the Walpole-Nornalup National Park (18 000 ha), much of which supported high open forests of karri, red tingle and yellow tingle. Three other national parks deserve mention: the Warren and Beedelup National Parks near Pemberton, together 4 500 ha in extent, are largely surrounded by State forests and contain high quality karri forest, and the Porongurup National Park, which is an eastern outlier from the main karri occurrence. There were also some small national parks and nature reserves, mostly surrounded by private property, and hence of limited long-term conservation value. This did not, however, detract from their importance for short-term conservation, recreation, landscape enhancement and other amenity purposes.

Public pressure for the creation of national parks and nature reserves as opposed to the more utilitarian State forests, began slightly earlier than the latter, and had a different history. In the 1890s a number of scientific societies were formed in Western Australia which numbered among their supporters prominent citizens, notably the Premier Sir John Forrest, the Governor Sir William Robinson, the Director of the Western Australian Museum Mr Bernard Woodward and the famous Victorian botanist Baron Sir Ferdinand von Mueller. Their influence led to the declaration of a large (64 000 ha) reserve in the northern jarrah forest east and north of Pinjarra in 1894. It did not, however, survive the wider pro-development attitudes of the times, and in 1911 the reserve was swallowed up in timber permits and eventually became State forest. Although these societies and their successors and others (Royal Society of W.A., W.A. Naturalists' Club, Gould League of W.A., Royal Australasian Ornithologists Union, the Royal Western Australian Historical Society, Tree Society, National Trust of Australia (W.A.), Anthropological

Society of W.A.) took no known initiatives in the nomination of any karri forest reserves, they were important in shaping public opinion in favour of conservation. Most of the earlier initiatives for reservation came from private individuals, local government and from officers of the Lands Department.

The Lands Act 1898 and the Permanent Reserves Act 1899 were important for conservation in that land could now be securely set aside specifically for the protection of indigenous flora and fauna, and provision made for their management by boards of control. The first karri forest national parks were all declared under these Acts: Walpole-Nornalup in 1910 and 1924; Warren 1909; Beedelup 1909; Porongurup 1925. Important nature reserves such as Lake Muir were also set aside.

The Country Water Supply section of the Public Works Department (later part of the Water Authority of W.A.) also has had a significant indirect influence on reservation for conservation. Protection of catchments by the gazettal of water reserves, usually but not always duplicating State forests, added substantially to the security of tenure of many areas of State forest. Long experience confirmed by research, has established the basic compatibility between water production and prudent sustained yield timber management. Recognition that native vegetation in catchments has great benefits for the quality of water produced therefrom has been valuable for conservation, because water is a forest value which receives universal support.

By the 1960s a substantial secure karri forest estate in public ownership was in existence. Most was in State forest, with a large scale timber industry actively dependent upon it. Conservation of karri in the 'preservation' sense was poor, being dependent largely upon a few, mainly small reserves. But the opportunity to alter that situation was available.

Conservation Reserves

Multiple use forest management

From its inception, forestry has accommodated uses other than timber production. Some of the earliest European forest reservations were fauna reserves which came about largely in response to the royal requirement to hunt deer. In Western Australia multiple use has always been practised according to the relative demands of society on the forests. Early on, the Forests Department accommodated recreation, botanical and zoological research, grazing, apiary and water catchment values, e.g. Working Plan No. 1, Mundaring (Stoate and Kessell 1926).

From the 1960s onwards, demands for forest values, both in absolute and relative terms, began to change rapidly owing to socio-economic reasons and the Government's attempts to accommodate them.

Formal management of the forests for multiple use in Western Australia did not come overnight. It evolved gradually over more than a decade. The concept was introduced in *General Working Plan No. 85* (Forests Department of Western Australia 1972), officially accepted as Government policy in 1976 and further developed in *General Working Plan No. 86* (Forests Department of Western Australia 1977).

General Working Plan No. 87 (Forests Department of Western Australia 1982) officially prescribed a system of zoning, in which each forest zone was allocated a priority use. Other uses were encouraged, permitted or forbidden depending upon the degree to which they conflicted with the priority use. An area allocated a priority use was called a Management Priority Area (MPA) for that particular use. This priority or dominant use system had its origin in an earlier move to concentrate silvicultural activity in high quality regrowth jarrah stands in high rainfall areas, the intention being to make optimum use of the silvicultural dollar.

Giving highly productive areas a production priority led to the notion of giving other areas of special value another appropriate priority. The Perup forest east of Manjimup, known to contain important populations of rare mammals, was the first to be so considered, and was given a fauna protection priority. This was the first of a series of Fauna and Flora Management Priority Areas (Christensen 1973, 1974) to be zoned throughout State forest.

The Forests Department was aware that no credible alternative system of reservation for nature conservation in jarrah or karri forests existed, or indeed could exist, outside State forest. The reason for this was that most of the jarrah and karri in public ownership was in State forest, which at that time had timber production as its primary purpose. With few exceptions, existing national parks and nature reserves in forested areas were small and failed to meet any recognized criteria for adequacy. The Forests Department therefore accepted the responsibility for all aspects of forest conservation. An essential component of this policy was the setting aside from commercial use of a proportion of State forest, adequate in recognized scientific terms, for the purpose of biological preservation. Areas thus set aside were to be called MPAs for Flora, Fauna and Landscape.

At the same time as the Forests Department was setting up its conservation system in State forests, the Environmental Protection Authority (EPA) was taking a separate initiative with the same end in mind (see later in this Chapter).

A third initiative important to the conservation of karri was also being taken at this time. A private proposal for a south coast (later to be called D'Entrecasteaux) national park was developed by a group of foresters and submitted to the appropriate authorities. This proposal added an essential coastal component to the karri reserve system which was largely absent from State forest (see later this Chapter).

The initiative of the Forests Department

During the 1970s the Forests Department set about selecting and demarcating a series of MPAs for flora, fauna and landscape in the southern forests. Criteria used for the selection of MPAs were derived from a report by the Australian Academy of Science (Anon. 1975) and the 'Specht Report' (Specht *et al.* 1974).

The Australian Academy of Science Report outlined the need to preserve genetic diversity within designated key communities which are representative of their biome. Other essential, as well as desirable, characteristics of reserves were also discussed. In particular, emphasis was put on the need for reserves of sufficient size in order to maintain diversity in the long term.

According to the Australian Academy of Science

> Areas of 4 000 ha appear to have a relatively good chance of achieving their conservation objectives, though areas larger than 20 000 ha may be necessary for macropods. Smaller areas, buffered within State forest or water reserves, would ensure the conservation of many plant communities and associated small animals.

Elsewhere in the Report, 50 000 ha is suggested as being desirable for eastern grey kangaroos (*Macropus giganteus*) in humid areas of south-eastern Australia.

These and other criteria outlined in the Report suggested that the ideal conservation reserve system should have the following characteristics:

· *Representativeness*

The reserve system should include examples of the key biological communities which it is intended to conserve, and these should contain the widest possible range of habitat, biological and geomorphological diversity.

· *Size and structure*

Reserves should be large (20 000 to 50 000 ha), circular or almost so in shape and have natural boundaries such as coastlines, watersheds, rivers, or geomorphological features. Smaller reserves should be buffered by State forests or other reserves.

· *Naturalness*

Reserves should be in a natural, undisturbed state, free of introduced species and pathogens.

· *Security*

The tenure and purpose of management of reserves should be secured by an Act of Parliament, requiring the agreement of that body to effect change.

· *Management*

Reserves should be managed and protected by an appropriately skilled and funded organization. They should contain, or comprise, a core area which would remain inviolate from disturbance other than that for approved research. The core should be surrounded by a buffer area in which certain uses, compatible with the continued existence of the populations to be conserved, could be permitted.

· *Replication*

Reserves should be replicated, and should be interconnected by corridors to allow the migration of flora and fauna.

The ideal is rarely attainable, but having the ideal provides a goal in a situation where compromise is inevitable. Thus, although genetic diversity is the key characteristic sought, there must be some compromise between diversity and security. Numerous small reserves generally achieve the highest diversity, fewer large reserves confer better security. Nevertheless, a reserve system with good diversity and reasonable security would be difficult to reject in favour of one with less diversity and great security.

The best criteria for reserve selection can only be as good as the scientific knowledge of the time. The results obtained from using any set of selection criteria for reserve selection is also dependent on the level of knowledge and the state of the land at the time of selection. The state of the land and the level of knowledge of the southern forests during the 1970s was as follows.

Naturalness

Fortunately, in the mid-1970s, more than half the karri forest was still in a virgin state. Timber operations up to that time had been concentrated mainly in the western half. The opportunity to demarcate large reserves in a natural state therefore still existed. Attention was focused mainly on virgin forests, even though the Academy of Science Report was quite explicit in not ruling out cut-over and regenerated forests for inclusion in conservation reserves. This attention on virgin forests meant that much of what was then selected now qualifies within the definition of wilderness as defined by the Australian Heritage Commission (see Lesslie *et al.* 1988).

Vegetation, soils and landform

The structure, density and floristic composition of the overstorey of all State forests was well defined and accurately mapped at that time. The Forests Department Aerial Photography Interpretation (API) type-maps used a structural classification system of plant communities which combined height/life form of dominant plants and the projective area of ground covered by the foliage of the dominant plants in the ecosystem. Within these structural subdivisions, species composition of the overstorey was used to define forest type (alliances) (Fig. 3).

Figure 3 also shows the variation in floristic composition, height, form, and density of karri forest actually mapped on API maps. The API maps contain much more detail than that deemed necessary in the Specht Report.

Other useful information on API maps included degree of fire damage, presence of unusual tree species, rock outcrops, rivers and streams, lakes, wetlands, reed swamps, sedge or heathlands and riverine vegetation communities.

The major plant associations described by Christensen *et al.* (1985) (see Chapter 2) were well known. Moreover, foresters with long experience in the karri forest were readily able to distinguish major understorey plant communities often dominated by single species, e.g. netic (*Bossiaea laidlawiana*) in the Donnelly River valley at the western end of the main karri occurrence; karri wattle (*Acacia pentadenia*) in the eastern and southern areas; hazel (*Trymalium spathulatum*) frequently on the sites where karri grows tallest; *Acacia urophylla* on the more fertile sites; and *Hovea elliptica* in mixed karri-marri stands.

The geology, landform and soils of the area were also well known (see Chapter 2) and the broad relationships between landform, soils, climate and karri occurrence were understood. Thus it was known that karri distribution is limited primarily by rainfall and that within this limit its distribution is determined by landform and soils. Outliers from the main karri range are restricted primarily to well drained moisture gaining sites (Churchill 1968; Bradshaw and Lush 1981).

Fauna

The Academy of Science and Specht Reports conceded that fauna was not taken directly into consideration because too little was known about it Australia-wide at that time. Their rationale was that fauna would automatically be conserved in habitats existing in major vegetation alliances and associations. The situation in the early 1970s in the karri forest was similar, so the same rationale applied.

However, enough research had been done to know that the karri forest could not compare with the drier jarrah and wandoo forests with regard to richness in vertebrate species. The Perup forest was known to contain viable populations of rare mammals, viz., the woylie (*Bettongia penicillata*), the tammar wallaby (*Macropus eugenii*), the numbat (*Myrmecobius fasciatus*), the western ringtail possum (*Pseudochierus occidentalis*) and the chuditch (*Dasyurus geoffroii*) (see Chapter 2). This area was one of the first to come under special attention in Forests Department multiple use planning (Christensen 1973, 1974). In this case, the rationale was reversed, in that fauna conservation was the nominated priority value, which in turn ensured the conservation of a range of jarrah, wandoo and swamp associations. None of these belonged to the high, open forest biome.

Reserve size

The Academy of Science Report put emphasis on reserve size. The main scientific justification for large size (20 000 to 50 000 ha) was for the retention of an adequate gene pool of large, mobile mammals, in this case macropods (kangaroos). There was no evidence that kangaroos were under any threat in State forests in Western Australia. However, there are other species, particularly carnivores such as the chuditch, which also require large reserves. Other arguments in favour of large reserve size were also considered. These included a cushion or bufffer against unplanned human impact, greater security for rare species, room for natural evolutionary processes to express themselves and greater protection for the genetic diversity present.

Perhaps the most telling argument was the fact that the opportunity to create large reserves still existed. Not to take this opportunity would have been an imprudent sacrifice of an option which may never occur again.

The choice of reserve areas

In the light of what was available, the level of scientific knowledge and the selection criteria based on the Academy of Science Report, the following approach to developing a reserve system in the karri forest was adopted:

Firstly, two large virgin forests were sought. It was hoped to find areas in the 30 000 to 50 000 ha size range, with a maximum of diversity within their boundaries.

Secondly, a series of smaller reserves, preferably virgin forest, were sought which sampled all the major river basins and the outlier occurrences not already within secure reserves.

The Forests Department API maps, overlain with known data on geology, landform, soils and understorey vegetation, formed the basis of the selection. When appropriately coloured, the API maps were ideal for comparing alternative proposals because diversity could be compared visually, thereby allowing the rapid elimination of the least satisfactory options.

When the best options were difficult to separate, additional criteria were used. Karri is perhaps best known for its beauty. The catchments of the Deep and Frankland Rivers have particular appeal in this regard. High granitic outcrops, deep river valleys and a wealth of high vegetation diversity combine to show karri at its magnificent best. Broken topography allows the viewer to stand back a little and so appreciate the tiered splendour of the forest.

Entire undisturbed subcatchments, the presence of known restricted species, e.g. *Eucalyptus ficifolia*, *E. brevistylis*, permanent streams and pools, swamps and lakes were other characteristics which helped to favour one option over another.

Size of reserve could be effectively increased by locating an MPA for flora, fauna and landscape adjacent to an existing or proposed national park or nature reserve. Valuable use was made of the Warren National Park and the proposed South Coast (D'Entrecasteaux) National Park in this way.

Severe fire damage was evident over large tracts of the virgin forest. As wildfire is a natural factor of the karri environment, fire damage in virgin stands is not important in a strictly biological sense: providing there is undamaged forest nearby: regeneration comes as a result and a new stand replaces the old. The visual qualities of the original healthy stand is lost,however, and is replaced for a human lifetime by a graveyard of gaunt standing dead trees which preside over a tangle of scrub, coppice and regrowth below. Since timber values steadily worsen as rot degrades the standing dead trees, it became Forests Department policy to seek out the worst of the fire-damaged stands for prompt utilization and regeneration before degradation progressed beyond commercial limits. This policy was not welcomed by the timber industry because the economics of logging healthy stands was always better. In seeking MPAs for flora, fauna and

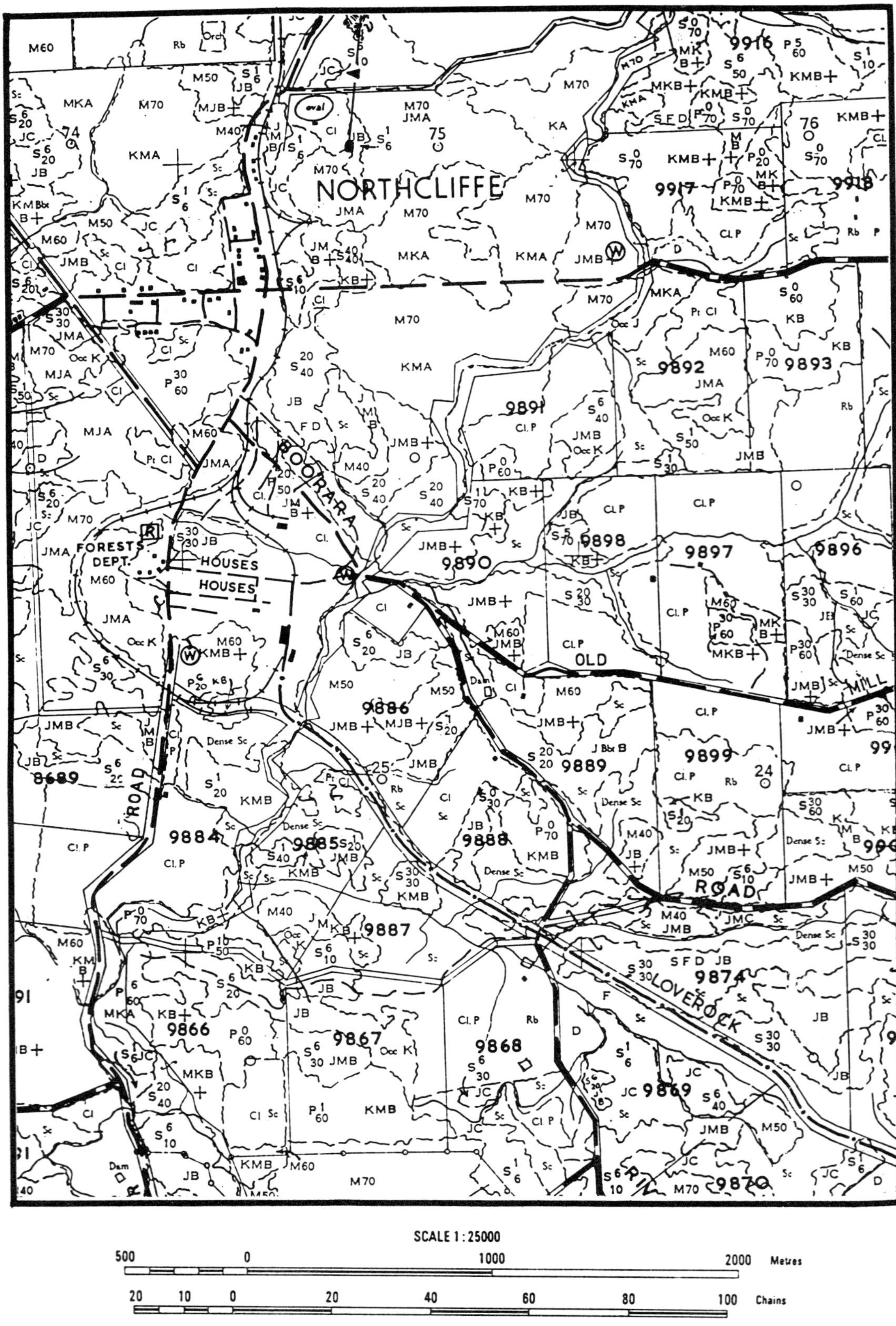

Figure 3
Section of an Air Photo Interpretation (API) map of an area near Northcliffe, to the north-west of Walpole.

LEGEND

1ST CLASS ROAD	
2ND CLASS ROAD	
3RD CLASS ROAD	
UNMAINTAINED TRACK	
REFERENCE TREE	THEODOLITE 2
	COMPASS 1
TRANSFERRED FROM TOPO SHEETS	4 3
PHOTO PRINCIPAL POINT	15
PHOTO PICTURE TIE	
PHOTO CONTROL POINT	G
PRIVATE ROAD – ACCESS RESTRICTED	

J	JARRAH
M	MARRI
K	KARRI
W	WANDOO
Bbt.	BLACKBUTT
Bull.	BULLICH
B'sia	BANKSIA
D	NON FOREST
D b	DIEBACK
P.D.	POSITION DOUBTFUL
F.D.	Light Fire Damage
S.F.D.	Severe Fire Damage

F	Flats
Sc	Scrub
Rb	Ringbarked
Orch.	Orchard
Pt. Cl.	Part Cleared
Cul.	Cultivated
P	Pasture
■	Building
S. T.	Shade Trees
V.O.	Very Open
Dec.	Decadent
V	Virgin Forest

FOREST ASSESSMENT

HEIGHT CLASSES

Metres

J M Bbt	
A+	30+
A	25-29
B+	20-24
B	15-19
C	under 15
KARRI	
A	50+
B	40-49
C	under 40

FOREST TYPE

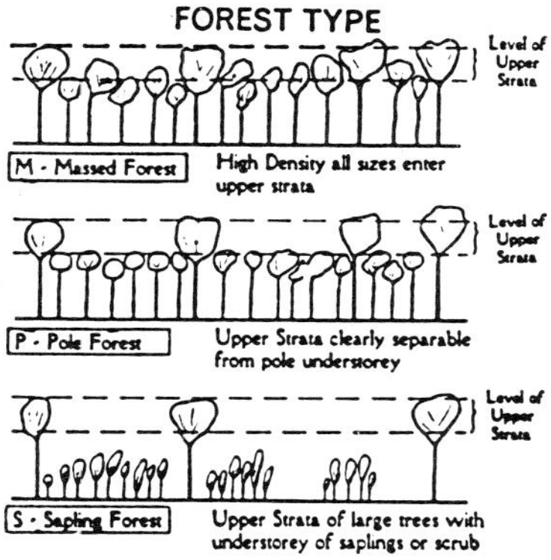

UPPER STRATA

Percentage Crown Cover

JARRAH		KARRI	
0	1 - 5	1 - 4	5 - 9
6 - 19	20 - 29	10 - 14	15 - 19
30 - 39	40 +	20 - 24	25 - 29
		30 - 34	35 - 39
			40 +

20=% Upper Strata
60=% Total Density
M type—total density only

landscape, all other things being equal, large tracts of severely fire-damaged forest were avoided, because their visual quality was judged to be impaired.

The first of the two very large reserves selected was the 30 000 ha Wattle-Soho area (which has since become the Mt. Frankland National Park), which straddles the upper catchments of the Weld, Deep and Frankland Rivers. This is an area of superb mature karri forest and great biological and landscape diversity.

The second large reserve selected, the Lower Shannon, was 50 000 ha, made up of the lower Shannon River basin, plus an adjacent segment of the proposed D'Entrecasteaux National Park. This is also an area of great beauty and diversity and encompasses coastal elements not found in the Wattle-Soho area

The smaller areas chosen to make up the karri reserve system were Curtin forest block (upper Shannon River), Boorara block (Gardner River), Dombakup and Hawke-Treen blocks (Warren River), Strickland block (Donnelly River), Chester block (Blackwood plateau), Boranup block (Leeuwin-Naturaliste Ridge), Giants block (Bow River), and Swarbrick, Keystone and Dawson blocks (Walpole River).

These conservation areas were supported by a network of recreation priority forests at One Tree Bridge (Donnelly River), Brockman (Warren River), Muirillup (Gardner River) and Mt. Frankland (Frankland River). Three of these are regenerated forests, two are virgin. According to the Academy of Science Report, all qualified as suitable for inclusion in conservation reserves. All have high incidental conservation value, though their management priority was for recreation.

The Forests Department was satisfied that this reserve system contained the genetic diversity necessary for scientific purposes, and captured the magic and majesty unique to mature virgin karri. Diversity exceeded that deemed necessary in the Academy of Science Report. Together with the proposed D'Entrecasteaux National Park and the adjacent Perup Fauna Priority Area, the view was taken that a very comprehensive reserve system had been developed. The proposals were put in place immediately.

Nevertheless, the system had a serious weakness. The strong tenure (State forest) was not matched by the security of its purpose, which could be changed without reference to Parliament. This was a matter of concern for the community and the EPA. The latter recommended that the purpose of management of MPAs for flora, fauna and landscape and MPAs for recreation should be given the status of a Parliamentary Regulation, which can be changed only by reference to Parliament, but this was not accepted by Government.

The initiative of the Environmental Protection Authority through its Conservation Through Reserves Committee

In December 1971 the EPA was established to, *inter alia,* 'consider and initiate the means of enhancing the quality of the environment'. The Authority recognized that the establishment of an adequate conservation reserve system would be an important mechanism for achieving this objective. Accordingly, it established the Conservation Through Reserves Committee (CTRC), which met for the first time in February 1972.

The CTRC divided the State of Western Australia into 12 natural areas called 'systems'. Systems 1 and 2 more or less covered the same area as that covered by the 'southern forests' of the Forests Department. The CTRC, in looking into reservation for conservation in Systems 1 and 2 acted independently, although the Forests Department was consulted.

In August 1974 the CTRC published its report to the EPA (CTRC 1974). This was then made available for public comment.

The CTRC took a different approach to that of the Forests Department in its consideration of reservation in the karri forest. Bound by its charter to seek opportunities for the creation of large reserves, the CTRC, after due consideration, chose the entire Shannon River basin. This, added to what was already reserved in national parks and nature reserves, met the aim of the CTRC: 'to set aside sufficient native habitat to be preserved and managed, both for the preservation of animals and plants, and for the enjoyment and education of the population' (CTRC 1974).

The CTRC wished to reserve the entire Shannon River watershed because it was a large single continuous area containing virgin karri and karri-marri forest. They proposed that this area would then act as a benchmark area for changes in biological, hydrological, pedological and sedimentological parameters which might take place elsewhere as a result of agriculture or timber harvesting. The presence of the Broke Inlet in a largely intact natural state at the mouth of the Shannon River, and the presence therein of a sedimentary delta of classic form, added to its value. The Shannon basin was seen as having the potential to become a reserve of world renown.

The initiative of a private group of people

Parallel with the development of the Forests Department's reservation system and the CTRC initiative, a third proposal emerged for the formation of a large south coast national park. This proposal was put forward by a group of foresters, Jack Bradshaw, Roger Underwood, Steve Quain, Don Grace and Barney White. Some of these were the same people developing the State forest reserves system. The south coast national park, later to be named 'D'Entrecasteaux', was officially proposed by the W.A. Division of the Institute of Foresters of Australia (Bradshaw *et al.* 1975). The Forests Department and the CTRC both gave the D'Entrecasteaux proposal their full support. The developers of the State forest reservation system were fully aware that important components of the wet sclerophyll existed in the proposed coastal park, and assumed, correctly, that these areas would be reserved in due course. The Lower Shannon Flora, Fauna and Landscape MPA depended upon its conjunction with the proposed national park to complete its diversity of landform and vegetation, and to round off its boundaries.

The close interdependence of 'high open' forest and rainfall means that karri seldom occurs far from the coast, and many of its associations are distinctly maritime. Hence, a system of reservation purporting to represent the wet sclerophyll forest in Western Australia has to include coastal landforms. The south coast area also had many of the qualities of wilderness (Lesslie *et al.* 1988).

The coming together of the three initiatives

The Forests Department did not endorse the CTRC Shannon River basin proposal. The main reasons for not doing so were, firstly, the belief that the Shannon River area failed to include important variations in the tall, open forest biome, and secondly, complications had arisen from the presence of severely fire-damaged forest in its mid catchment region. The Forests Department had spent years persuading a reluctant timber industry to utilize this fire-damaged timber so that effective regeneration could begin, and was reluctant to forego its recent hard-won success. Furthermore, the Forests Department reserve system already included two MPAs for flora, fauna and landscape in the Shannon River basin, one (Curtin) in the north, and a larger one (Lower Shannon), in the south.

The Forests Department was reluctant to accept the rest of the Shannon basin as a further addition to its reserve system of MPAs, which it believed was fully representative of biological and wilderness values. This was because it was not possible at that time to accommodate the loss of timber resource without severe disruption to the timber industry, nor was this considered to be justified.

When released for public comment, the CTRC report generated intense public opposition in some areas, and equally intense support in others. The EPA appointed a special review committee to review the CTRC recommendations and the public response to them.

The Special Review Committee (EPA 1976) preferred the Forests Department's approach to reservation in the karri forest and recommended its adoption.

The EPA's response to the CTRC report and the Special Review Committee (1976) was published in 1976 in its official recommendations to Government (EPA 1976). This has commonly been called the 'Red Book'. While recommending that the Forests Department reserve system be adopted, the EPA also recommended that logging and regeneration in the Shannon River basin proceed only in severely fire-damaged areas for a further five years, with the proviso that operations be limited to 9 per cent of the Shannon basin area. Thus the option to create a park at a later date was retained. Other recommendations were that MPAs for flora, fauna and landscape and MPAs for recreation be called 'forest parks', the management plans of which should receive the protection of a Parliamentary Regulation. In other words, parliamentary approval was required to effect change. The Red Book also recommended creation of the South Coast (D'Entrecasteaux) National Park (the area which had been proposed by the Institute of Foresters). All Red Book recommendations were endorsed by State Cabinet in October 1976.

Five years later the question of conservation in the karri forest was raised again by the EPA. In its submission to the EPA (Bradshaw and Lush 1981), the Forests Department reaffirmed its belief that its present reserve system was the best approach, and put conservation into the context of management of the karri forest as a whole.

To review the Forests Department submission and the responses to it from the public and the environmental movement (Conservation Council of W.A. 1982), the EPA obtained the services of an independent expert, Dr P Attiwill, Reader in Botany at the University of Melbourne.

The Attiwill Report (Attiwill 1982) also favoured the Forests Department approach, and expressed the belief that it was adequate for conservation in biological terms. The EPA reaffirmed that its 1976 recommendations were correct and should be fully implemented (EPA 1982), citing the Attiwill Report as being the rationale for its stance. The questions of greater security of purpose for areas with value to conservation, better nomenclature of reserves, management and the future of certain areas were among other important issues raised in the EPA Report. The EPA's position on the Shannon River basin was essentially the same as that in 1976, viz., that harvesting and regeneration of fire damaged forest should continue in the central basin with a view to its value as a conservation reserve being reviewed some time in the future.

In August 1983 a new Labour Government was elected, and immediately acted on a pre-election promise that the Shannon River basin would become a national park. The Forests Department was instructed to amend its *General Working Plan No. 87* (Forests Department of W.A. 1982) to make the entire Shannon River basin an MPA for flora, fauna and landscape.

This action thereby combined two systems of reservation in the karri forest, both made by competent bodies, and both believing that their respective approaches were adequate in biological terms.

The newly-elected Government also expedited the declaration of some portions of the proposed D'Entrecasteaux (South Coast) National Park, and set about to rationalize its agencies responsible for the management of natural lands under public ownership. The Department of Conservation and Land Management (CALM) was created in 1985. It was formed from an amalgamation of the Forests Department, the National Parks Authority and the Wildlife section of the Department of Fisheries and Wildlife. CALM became the management agency servicing the newly-formed National Parks and Nature Conservation Authority (NPNCA), and the Lands and Forest Commission (LFC). National parks and nature reserves (marine and terrestrial) were vested in the former, and State forests were vested in the latter.

In April 1987, CALM published for public comment, draft management plans for the Southern Forest Region (see Table 2b).

In August 1987, following concerns raised in Parliament on the Government's proposal to effect its 1983 decision to declare the Shannon River basin a national park, an expert Committee of Inquiry was established.

The terms of reference for the Committee of Inquiry were to:

- examine the history of the reserve proposal;
- consider its validity in light of the current knowledge on southern forests ecosystems and the proposals in the then draft regional management plans; and
- consider the impact of the reservation on the Government's capacity to meet its timber supply commitments.

The Committee concluded that the reservation had scientific merit and could be made without any derogation of previous reserve proposals or reduction in the level of economic activity in the timber industry (Committee of Inquiry on the Shannon River Basin 1987).

In December 1987, following a period of public comment and review, the final management plans for the Central, Southern and Northern Forest Regions were approved by the Government and published, thereby becoming the official blueprint for the ensuing 10 years (CALM 1987a, 1987b and 1987c). These replaced the *General Working Plan No. 87* (Forests Department of Western Australia 1982) and any management plans extant for national parks or nature reserves within the areas. All State forests, national parks and nature reserves were brought under one management.

Many changes to tenure were proposed. Additional national parks and other reserves were created. Some boundaries were adjusted. In the southern forests all previous MPAs for flora, fauna and landscape and MPAs for recreation were scheduled to become either national park, nature reserve or to be put into a new category called 'conservation park'. All were to have 'A' Class security of tenure and purpose which requires approval of both Houses of Parliament to effect any change. The weakness of the old Forests Department's MPA system was therefore finally overcome.

The system of reservation in the southern forests as of February 1988 is that shown in the management plans for the Central and Southern Forest Regions (CALM 1987a, 1987b), and is reproduced in Figure 4 and Appendix 2 of this book. To obtain an overview, the above management plans should be read in conjunction with the management plan for the combined Shannon National Park and D'Entrecasteaux National Park adopted in December 1987 (CALM 1987d).

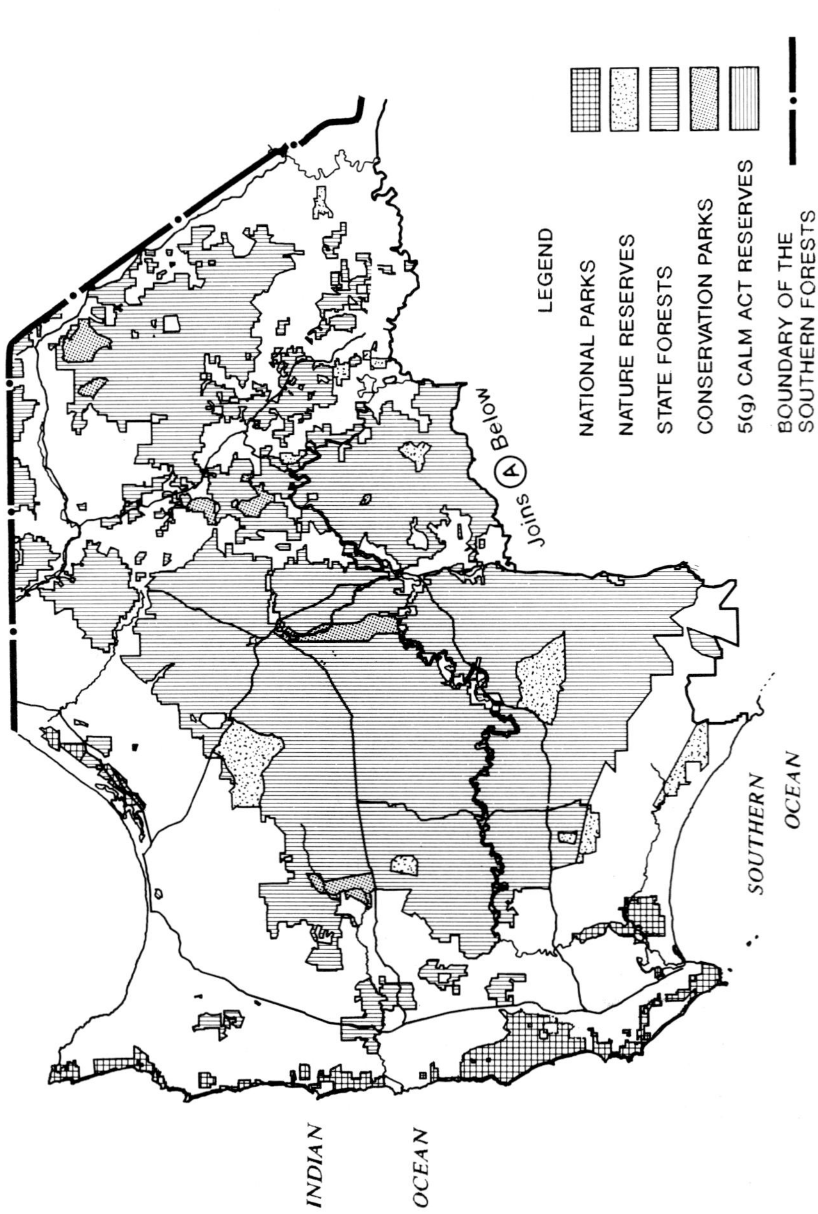
LEGEND
NATIONAL PARKS
NATURE RESERVES
STATE FORESTS
CONSERVATION PARKS
5(g) CALM ACT RESERVES
BOUNDARY OF THE SOUTHERN FORESTS
Joins A Below
INDIAN OCEAN
SOUTHERN OCEAN

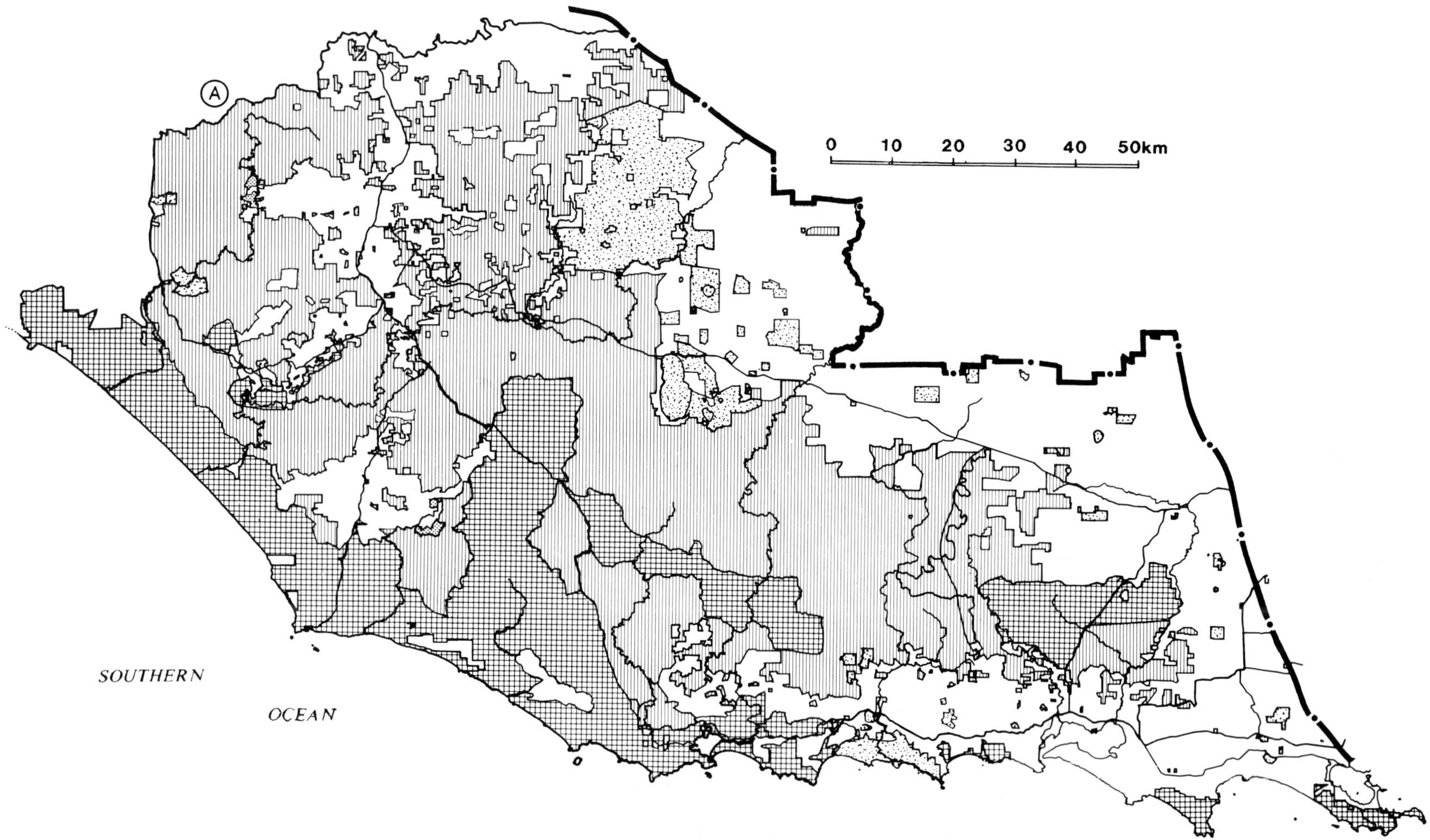

Figure 4
Map of national parks and nature reserves in the southern forests (as proposed by CALM 1987b).

During 1991 the forest reserve system was again reviewed. This followed a detailed joint review of heritage values in the southern forests between CALM and the Australian Heritage Commission. Revised forest management plans will be released in draft form for public review and comment in 1992.

Summary

The present reserve system is an amalgam of different selection processes by different groups at different times. Conservation of plant and animal communities is a comparatively recent phenomenon. At the time of European settlement in Western Australia there was little recognition of the need for nature conservation. The early dedication of national parks at the beginning of the century was based as much on aesthetic appeal as on conservation in the modern sense.

Later reserve selections were based on modern scientific principles, including both biological and other values. It is an enormous strength of the system that it has been selected in this way, by different input from a very wide range of people both recently and further back in time. All used different selection criteria based on their background and the socio-economic and environmental values of the time.

Each generation tends to think of their values as the only ones that are valid and worthwhile, often giving scant regard to the values of their forebears. However, values change, and who is to say which values are the most significant or the most correct. It is worth reflecting that it was recognition by an earlier generation of the value of our forests for timber, that we have to thank today for the largely unspoilt forests and wilderness areas which survive.

It should be said that the selection of the current reserve system did not take account of heritage values or wilderness in any formal sense. Objective criteria are now available for evaluating wilderness, and the opportunity is there to consider the adequacy of the reserve system with regard to these values. CALM is in this process of doing this.

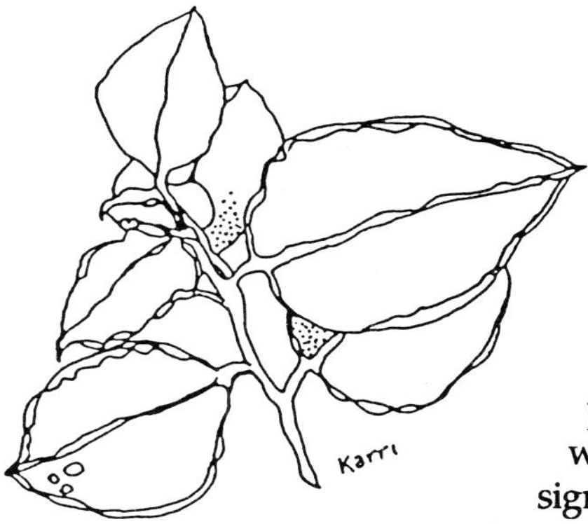

CHAPTER 6

MULTIPLE USE FOREST MANAGEMENT AND CONSERVATION

Introduction

The transient effects on the forest of timber production and regeneration operations (as opposed to some other primary production, e.g. clearing for farming) means that forests which are used for the production of timber also have very significant nature conservation value (see also Recher 1986; Davey 1990). Their value, especially for wildlife, is greatly influenced by the way in which they are managed (Recher *et al.* 1987; Davey 1990).

The many aspects of forest management which influence the conservation values of multiple use forests are discussed in this chapter. Karri forest management is dealt with in particular, as the clearfelling method of harvesting and regenerating which is used in the karri forest, is poorly understood in most sections of the community. As in the previous chapter, an historical perspective is taken to give an understanding of the underlying forces which with time have influenced change. As with conservation, political and social factors have been equally as important as science in determining the way the forest is managed.

Silviculture

Silvicultural characteristics of karri

Silviculture is the branch of forestry that deals with the establishment and tending of forests. It is the basis of forest management: the very essence of forestry. The sort of silviculture which is practised depends on the objectives of management, the species of tree, its life form, growth habit and ecology, together with other parameters of the forest ecosystem, as well as socio-economic factors.

Karri is a high open eucalypt forest species which has characteristics in common with similar forest eucalypts in other areas of Australia (*E. regnans, E. delegatensis, E. fastigata, E. nitens, E. grandis, E. deglupta*). All exceed 45 m in average height. All require high annual rainfall, freedom from prolonged summer drought and deep well drained soils. None develops lignotubers, all being dependent upon seed for regeneration. All grow in forests with dense undergrowth, which inhibits the survival of seedlings unless subject to heavy disturbance. Large gaps in the canopy equal to at least one tree height in diameter are required for regeneration to develop rapidly into mature trees. High intensity wildfire is the usual natural means of creating broadscale conditions suitable for regeneration and subsequent development of the young trees. All species mentioned above have a tendency to occur naturally in even-aged stands or patches.

Forest managers have long recognized the clearfelling system of silviculture as the most effective and economically efficient way of harvesting and regenerating these forests (Jacobs 1955; Hillis and Brown 1978). However, the alternative system, selection cutting and regeneration has also been used in the karri forest.

A description of these two systems of harvesting and regenerating is appropriate here before describing the history of the development of silvicultural systems in the karri forest.

In Western Australia, clearfelling and regeneration is regarded by foresters as the cutting of all trees on an area of not less than 10 ha followed by a regeneration burn. Regeneration is achieved through seed from seed trees which are left uncut within the area for the purpose or by planting with karri plants raised in the nursery. Where seed trees are used these trees are harvested immediately regeneration has been achieved. Clearfelling and regeneration is easy, cheap, highly successful, disturbs a small area of forest, and produces even-aged regrowth stands well suited to ongoing management for wood production. Rate of regrowth is fast, thinning is easy to program and control, inventory is simple, and fire protection by prescribed burning to reduce fuel is feasible. Against all this, the immediate impact of clearfelling on forest structure is considerable, even though disturbance of this magnitude occurs only once per rotation of 100 years or more (intermediate thinnings require only light machinery and no intense regeneration burn). The appearance of a clearfelled coupe after the regeneration burn is one of devastation. However, karri regeneration is fast, and from about 15-years-old onward the forest 'feel' returns and progressively improves.

'Group selection' cutting and regeneration in the karri forest, when it was in practice, involved cutting small groups of trees, generally not more than two hectares followed by regeneration burning. Regeneration was achieved by seed from the remaining standing trees. The system is more difficult and expensive to practise than clearfelling, and more prone to regeneration failure, particularly if the regeneration burn exceeds or falls short of prescribed intensity. Damage to the standing trees by the regeneration fire is also difficult to avoid. Effective growth in the mixed-aged stands produced can be low, thinnings and inventory are difficult, and prescribed burning for fuel reduction is difficult and risky. In terms of soil disturbance and change to forest structure, however, the impact is less, though still significant because large machinery has still to be used (Bradshaw and Lush 1981; Breidahl and Hewett 1992).

A similar harvesting operation using big machinery followed by a high intensity regeneration burn must be repeated after 30 years or so (intermediate thinnings would occur in similar patterns to clearfelled stands). Selectively cut stands have a much better appearance because retained trees still dominate the landscape. These trees also provide nesting places, which are not provided by young stands after clearfelling, for some birds and mammals. In a forest subject to clearfelling, hole-nesting species in particular must be catered for by providing an adequate and well distributed system of uncut reserves and forest corridors. Successfully managed selection systems can mask the immediate effect of tree harvests with the result that the forest 'atmosphere' is never entirely lost.

Clearfelling is the more appropriate system to use in areas where efficient and economic timber production is important. Selection cutting may be more appropriate in 'special' cases such as special care areas adjacent to steep slopes, surrounds of rock outcrops, swamps, or groves of trees having special recreational or nature conservation value.

The development of karri silvicultural systems

Prior to the passing of the Forests Act in 1918, and the formation of the Forests Department in 1919, unregulated cutting in karri amounted to clearfelling. Wildfires burnt through the cut-over areas, and seed from unmarketable 'cull' trees was released to germinate in the ashes. Fine stands of regrowth resulted. Remnants of such stands can be seen on farms in the Denmark-Torbay area where karri forest was made available for agriculture following timber cutting. At Boranup near Karridale, the history following cutting was different. The limestone soils were not favourable for agriculture and the land remained under State control with the result that a tract of regrowth forest about 100 years old is now in existence. Most who are unaware of its history do not distinguish between this and virgin forest (Plate 20). The stand is now incorporated into the Leeuwin-Naturaliste National Park.

After 1919 and the formation of the Forests Department, areas of karri that were either dedicated State forest or mooted for such were systematically regenerated by the Forests Department following timber felling. The procedure used was a clearfelling system that used cull trees left standing as a seed source. Unwanted culls, marri and understorey species were felled following the commercial operation. A regeneration burn was programmed when ample seed existed in the crowns of the culls, which were then ringbarked following the completion of their seed tree function. Most of the beautiful regrowth forests of Big Brook and Treen Brook near Pemberton were regenerated by the application of this system (Plate 21).

Though the clear felling and regeneration system used in the 1920s and 1930s was successful, changes were introduced in the late 1930s.

Then, as now, people unfamiliar with natural processes found it difficult to look upon clearfelled areas as other than forest destruction and there was pressure to release such areas for agriculture.

Selection felling and regeneration which left trees standing in cut-over areas of forest helped the Forests Department officers to advocate its retention as State forest in the ongoing agriculture-forestry land use conflict.

A gradual change to a selection felling system and regeneration was occurring, which by 1938 had become officially recognized by the Forests Department.

As the selection cut and regenerated stands developed, and difficulties in management surfaced, professional opinion began to change back in favour of clearfelling and even-aged management. Perhaps the most telling evidence in its favour was the superiority in health and vigour of the extensive even-aged Big Brook and Treen Brook stands over those being produced by the selection system. In those times, burning for regeneration in partially cut stands was far less well understood than it is today. As a result, fire intensity could be either too high, resulting in good regeneration but damage to the retained trees, or too low, resulting in healthy retained trees but poor regeneration permanently inhibited by scrub. Growth of saplings was disappointing, being inhibited by mature trees nearby (see Rotheram 1982). The presence of young, fire-sensitive regrowth in mixed-aged stands prevented fuel reduction burning, thereby creating an increasing fire protection problem. A further problem was that logging technology of the time could not prevent extensive damage to regrowth when the selection-cut stands were harvested a second time. This still remains the case.

Following large-scale field trials, clearfelling was reintroduced into karri operations in 1968. The new system relied upon the retention of selected seed trees, the best trees in the stand, which were later removed after completing their function. Natural seedfall was supplemented by nursery-raised open-rooted seedlings which could be used in non-seed years and thus avoid creating a backlog of unregenerated cut-over forest area. Research work in the meantime had clarified the seed development cycle, allowing prediction of seedfall to be made (Loneragan 1979), and demonstrated the feasibility of using cheap open-rooted karri nursery stock for planting in the field. Spot application of fertilizer at the time of planting guaranteed good, even, early growth and high survival (see Breidahl and Hewett 1992). Regeneration success year after year was assured.

In 1975, seven years after the reintroduction of the clearfelling system, the marri woodchip industry commenced. This utilized both marri and karri logs of subsawmill standard. For the first time the utilization and proper regeneration of mixed karri-marri stands became feasible. Previously it had not been possible to utilize marri efficiently because most of the stems were unsuitable for sawmilling. So many more logs per unit area of forest were now available that the effective sawlog yield and hence the overall karri sawlog resource, increased by some 15 to 25 per cent. The advent of woodchipping also meant that thinning of regrowth stands established many years earlier also became economically feasible.

In areas where karri occurs approximately half of its occurrence is with other species, especially marri. Prior to 1970 karri harvesting and regeneration operations were confined largely to stands of pure karri. At the time there was no market for marri and so the trees had to be left standing and it was difficult to properly regenerate the karri component of the stands with the marri still present (see Breidahl and Hewett 1992). Following the establishment of the marri chipwood industry marri trees became commercially valuable and it was then possible to harvest and regenerate the mixed stands.

State forest was not the only beneficiary of the woodchipping industry. Stands of regrowth karri and/or marri on private property became valuable almost overnight, and hence could be viewed by their owners as permanent income generators in their own right rather than areas to be converted to pasture or crop as soon as possible. Such stands conserved by farmers for wood production are a gain for conservation.

In summary, karri forest has been harvested and regenerated in large-scale operations for over 100 years, and a very large amount of research has been done on karri silviculture (see Breidahl and Hewett 1992). Two basically different silvicultural systems, clearfelling and selection cutting and regeneration, have each been used for long periods during that time. In evaluating the two methods, foresters are in the fortunate position of having long experience and thousands of hectares of results to call upon. Both systems can be made to work. Indeed, one grades into the other (Bradshaw 1992 in press).

Water

In much of the south-west soil-stored salt threatens the quality of the available water. Clearing for agriculture in the upper catchments of the major streams flowing to the coast has salinized streamflow which was fresh prior to the removal of natural vegetation. Only those short streams which rise mainly in the forested area have remained fresh (see Schofield *et al.* 1988; Anon. 1989).

The main concerns in southern forests are the possible effects on stream salinity, sediment load and streamflow volume, of intensive timber harvesting operations.

The long history of intensive timber harvesting in the karri forest, with no obvious associated degradation in water quality, suggested that there was little conflict between the two. Nevertheless, the importance of fresh streamflow and the greater intensity of cutting anticipated by the introduction of a woodchip industry in the mid 1970s, induced the EPA to take steps to investigate the question. A steering committee was formed to guide research to monitor water quality, identify areas prone to stream salinization and to provide the Forests Department with data on which to base management. As a result, 11 research catchments which spanned the rainfall zones and forest types of the area were instrumented in 1975-76; four of these were immediately logged and regenerated as early warning indicators of any potentially serious environmental effects. The remainder were paired catchments with untreated controls intended for long-term investigation.

Analysis of the results of 10 years of monitoring indicates that logging and regeneration methods involving clearfelling of karri and heavy selection cutting of jarrah have only very minor and transient effects on stream salinity and sediment concentrations. No major changes to management practices were recommended, only some refinements to current practice to moderate transient effects on stream salinity and sediment concentrations (Anon. 1980; Anon. 1987a; Borg *et al.* 1987a; Borg *et al.* 1987b; Stoneman *et al.* 1988).

In the high rainfall zone (>1100 mm per annum), where most karri occurs, the recommendations focused on measures to improve sediment control. Several options are available, including:

- redistribution of existing undisturbed road, river and stream zones to increase zones along streamlines;
- exclusion of forest operations likely to cause soil and understorey disturbance from an undisturbed buffer zone along all streamlines for the period of up-slope logging operations, and for at least three years following up-slope regeneration (termed 'phased logging');
- improvement in scheduling of logging and road construction to further limit winter operations in the most sensitive areas; and
- further improvement in the location, construction and maintenance of logging roads.

The above recommendations confirm experience worldwide that undisturbed stream-side vegetation is effective in controlling sedimentation, that winter logging poses most risks, and that roads are the single largest contributor to sedimentation. These options are currently under review, or have already been implemented.

In the intermediate rainfall zone (900 to 1100 mm), saline groundwater is sufficiently close to the surface to contribute to streamflow before logging, and the post-logging rise in water table was reflected in increased salinity in streamflow, particularly early after its seasonal commencement. However, although flow-weighted stream salinity increased by 50 to 150 mg L^{-1}, it still remained below 500 mg L^{-1} (a recognized human consumption limit); salinity levels are declining and are expected to return to pre-logging levels as regrowth regains typical forest structure. Noteworthy salinity increases occurred only where pre-logging groundwater levels were close to the surface and where no stream buffers were left. Location of sensitive areas in advance and retention of buffers therein should control the problem in the intermediate rainfall zone. Some karri forest occurs in this zone, but it is mainly covered with jarrah forest, i.e. areas selectively harvested, not clearfelled.

In the low rainfall zone (<900 mm per annum), groundwater levels were generally low enough (9 m) not to contribute to streamflow pre or post-logging (Anon. 1980; Anon. 1987a; Borg *et al.* 1987a, 1987b; Martin 1987). The rapidity with which regeneration grows and evapotranspiration is restored to pre-logging levels mean that groundwater is most unlikely to contribute to streamflow. Long periods (i.e. decades) devoid of trees and other perennial natural vegetation cover would be needed for this to occur. As a result, integrated sawlog and woodchip logging in the low rainfall zone is not expected to generate salinity, so long as care is exercised in locating salt-sensitive spots and buffers are retained in these areas (Anon. 1987a).

Intensive harvesting operations increased streamflow yields for two or three years after logging, after which yields tended to reduce towards pre-logging levels. Vigorously regrowing stands, particularly in the high rainfall zone, could reduce yields below mature stand levels, unless stand density is modified by thinning as is planned (Anon. 1987b, 1987c).

Soil

Being the medium for root growth, soil is a vital part of all ecosystems. Damage to soil means damage to the ecosystem, and must therefore be prevented.

Forest soil can be degraded physically by uncontrolled logging or recreational use, or biologically or chemically by fire. In logging, soil damage is potentially caused by skidding operations (i.e. movement of the log from stump to landing). Highest risks occur when the soil becomes saturated with water (Wronski 1984).

CALM's strategy is to stop logging once a critical level of soil moisture is reached. In practice, this level is best identified by watching the machines at work and stopping them once the first suggestion of soil damage occurs. The potential for soil damage varies with soil type, season and type of operation. Procedures are set out in CALM's 'Manual of Logging Specifications for management of hardwood and softwood logging operations in the south-west of W.A.' (CALM 1990b). Recommencement of operations is dependent upon a defined Soil Dryness Index being reached (CALM 1990b).

Erosion is controlled by installing interceptor banks and drains at regular spacings across all snig tracks. Spacing varies with slope and soil type (CALM 1990b). Work on drainage begins when operations commence in a coupe, or when rain is imminent.

Soil compaction is the most difficult soil damage to rehabilitate, and it is kept to a minimum through the use of the 'Manual of Logging Specifications'. In log loading areas some compaction is unavoidable. All log loading areas (called 'landings') are rehabilitated by levelling and contour ripping to a depth of 500 mm and spacing of 1 m. In karri areas, landings and major snig tracks are replanted to karri and sown with seed of ground cover species.

Most soil 'damage' is soil disturbance rather than compaction, i.e. the scarification of surface soil, or its movement a few metres from its original position. Soil disturbance is inevitable in logging, and is often beneficial to regeneration. To date in the karri forest it appears to be of little environmental consequence.

Soil damage potentially caused by recreational activity in the forest is prevented on a case by case basis. Some groves of trees have been fenced off and paths made to direct and confine visitors. In some places raised walkways have been erected to avoid damage to sensitive, usually wet, areas.

Flora, Fauna and Landscape

The system used to conserve flora and fauna in southern forests subject to timber cutting was developed in the early 1970s in parallel with the development of the reserve system described in Chapter 5. Although biological research had been in progress in the karri forest for some years, insufficient results were then available to provide a completely objective basis for procedures. The importance of habitat trees as nesting sites and general habitat for some species was known, and hence the basic problem was their provision in regrowth forests. Other important issues were the retention of uncommon and restricted communities and ecosystems, and fire management.

Two realistic options were available for the provision of habitat trees: they could be left scattered through the new regrowth, or they could be left within strips or patches of forest left uncut. Other requirements, such as the need to protect streamflow, amenity and recreational values were also taken into account.

Leaving patches or strips of uncut forest was the strategy adopted at the time because the biological benefits were more certain, e.g. they could be located along the streamlines which were known to be the most important habitats, and to contain the greatest diversity and abundance of fauna, and they could act as uncut corridors linking to MPAs for flora, fauna and landscape. The problem of depressed growth to the new crop from competition from overmature trees scattered throughout was avoided, as was the fire control problem of having standing dead trees within regrowth stands.

Uncut strips were located along major streams and the most heavily used roads. Strip width relative to length was considered. The result was a tendency to protect fewer of the bigger streams with wide uncut zones, and to have wide strips reserved along the most used roads. Experimental map designs of typical forest blocks suggested that about 20 per cent of their area was a good figure, and this was accepted as a guideline for all forest blocks (these are management units of 2 000 to 5 000 ha).

Formal adoption of the road, river and stream buffering system in the mid-1970s (Forests Department of Western Australia 1977) served a number of other purposes. River and stream zones were known through the scientific literature to protect watercourses. Road zones maintained a vista of undisturbed forest for travellers through the area, and also acted as linking corridors between reserves. Selected major roads were given a reserve width of 400 m either side (total 800 m); major rivers like the Donnelly and the Warren had zones of 200 m either side (total 400 m); most third order streams and 30 per cent of second order streams had zones of 100 m either side (total 200 m) (see Fig. 5a).

Zones were deliberately linked to other zones, e.g. stream zones were often extended to interconnect across saddles and ridges. In addition, 'special care' zones were established where there was no stream zone but where slopes exceeded 15°, or in forest surrounding wetlands, rock outcrops, historic or recreational sites, and research areas. These zones were also retained uncut.

When the road, river and stream zone system was established, planners were aware that it had some shortcomings. It was obvious, for example, that to preserve mature forest views along roads, the actual uncut width required would vary with forest type, topography, road alignment and subjective judgement of what constituted good viewing. At the time, the Forests Department had neither the expertise nor the resources to solve these problems and therefore postponed making a final, irrevocable decision by adopting a conservative stance which involved doubling the originally estimated 200 m required to 400 m either side. This effectively preserved the option to deal more effectively with the problem at a later date. Selecting streams to be protected posed a similar problem. It was fully expected that changes would be required with increasing experience and research.

In the 15 years since the establishment of the road, river and stream zones, CALM (previously the Forests Department) has acquired wider expertise in landscape design and much more research has been done in hydrology, biology and silviculture. Digitized and computerized management systems have been introduced, and a review of the original system is currently underway (see Fig. 5b).

Conservation at the regional and species levels

No vertebrate species are considered to be in danger of extinction as a result of current harvesting and regeneration practice in the southern forests of Western Australia. The existing reserve system and proposed long harvesting rotations ensures their future. At issue is whether existing species should be conserved at all times at a local level. It should be noted that the level of knowledge of invertebrates is such that the same cannot be said with confidence for this very large group. It is questionable whether we will ever know the full cycle of extinction and evolution of new species even in the pristine natural forest environment.

To assist in conserving species at the local level, an intricate system of road, river and stream zones totalling at least 20 per cent of <u>every forest block</u> is left uncut. This area totals 76 100 ha in southern forests.

In addition to the conservation reserve system and the road river and stream zone system, there are other areas which have been identified and excluded from harvesting, e.g. heathland,

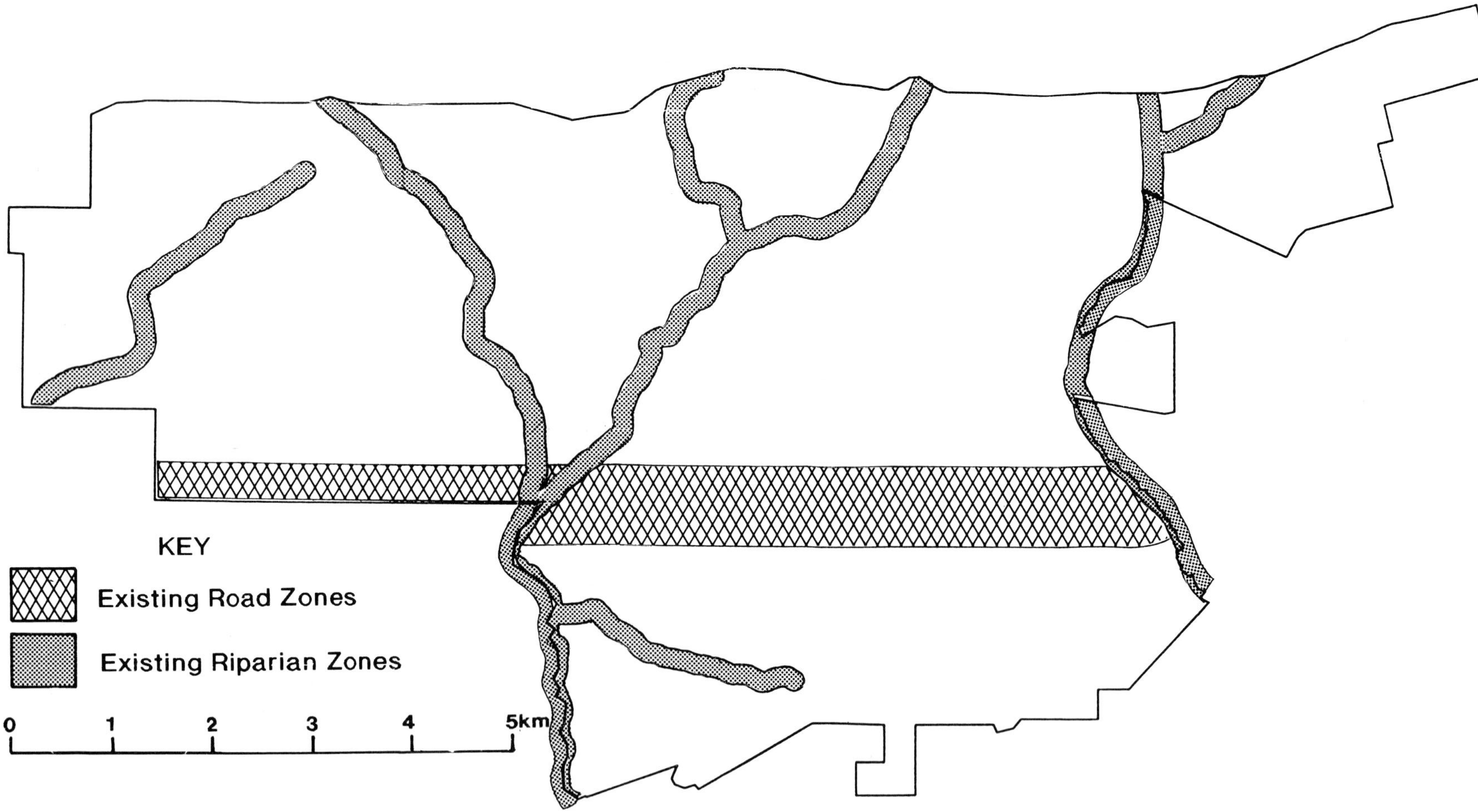

Figure 5a
Example of the existing road, river and stream zone system in a portion of one forest management block, Jane Block, near Pemberton.

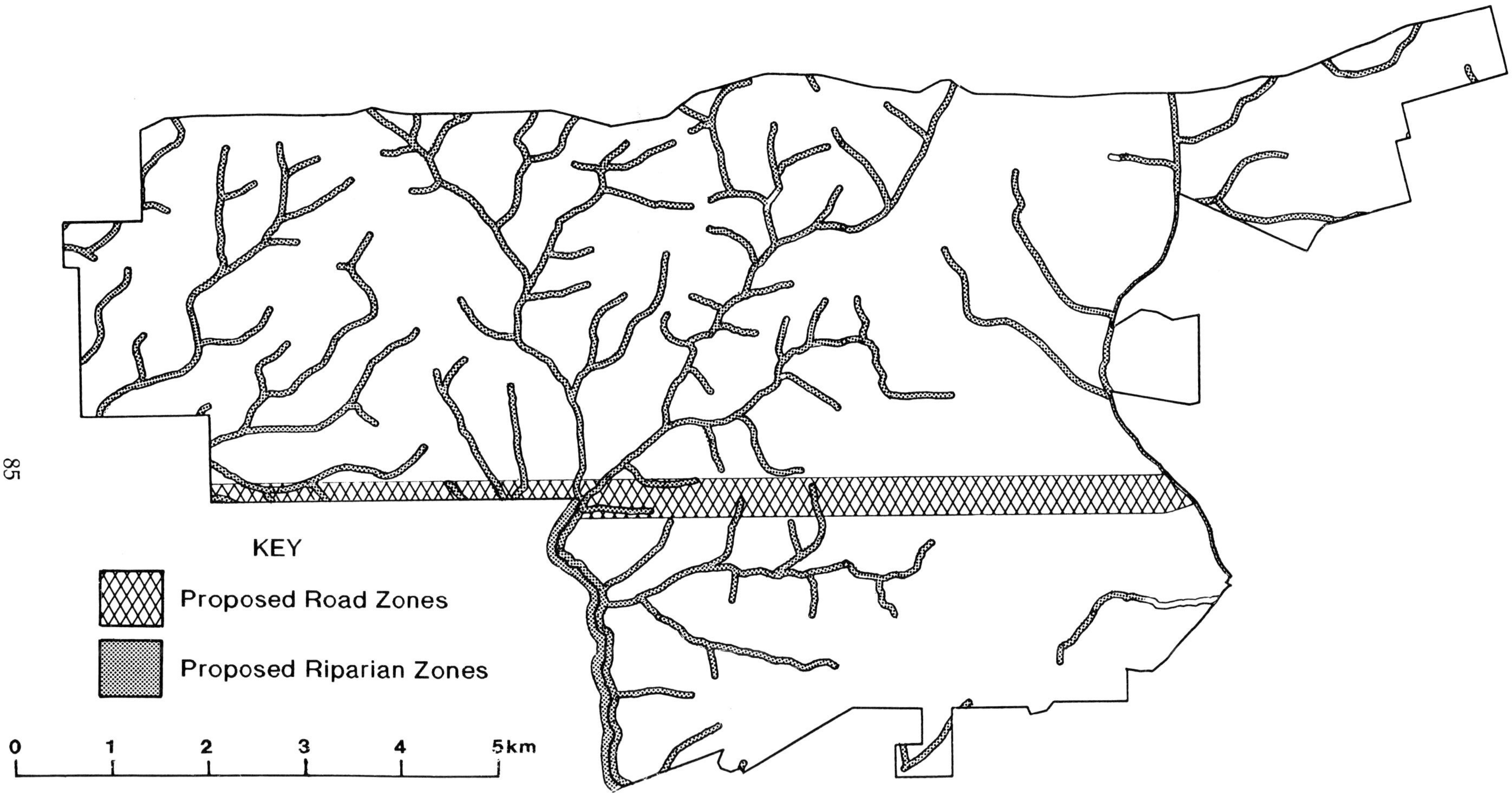

Figure 5b
Example of the proposed road, river and stream zone system in a portion of one forest management block, the same area as depicted in Figure 5a.

sedgeland, woodland and rock outcrops. These total some 143 100 ha in the southern forests (see Table 4) and have particularly high conservation value.

The network of uncut road, river and stream zones, together with the heathland, sedgeland, woodland and rock outcrop reserves and the national parks, nature reserves and conservation parks, form a complex network of mature uncut forest throughout the southern forests. This area, together with the developing regrowth forest, which forms a mosaic throughout the area, provides areas from which animals can recolonize regenerated stands. The young of all species have to disperse, and this is the basis of the colonization process described. The zones and reserves also provide hollow trees in which the hole-nesting species living and feeding in regenerated forest can nest.

The areas which are logged recover with time; felling only creates a temporary disturbance to the forest ecosystem. The degree of disturbance and the time it takes for the forest to return to the pre-disturbance situation depends very largely on the felling and regeneration system (clearfelling or selection felling) used, the type of forest, and the rotation length adopted. Although the detail for each situation has yet to be documented, the overall picture is clear. In their study on classification and evaluation of sites in karri regeneration using floristic attributes, Inions *et al.* (1990b) demonstrate a strong relationship between floristic pattern and edaphic and climatic variables. Following clearfelling and burning, community type as defined by a series of indicator species remained unchanged in 143 out of 144 plots. The exception occurred in a quadrat on a log landing. They conclude that clearfelling and fire regimes are not a primary determinant of perennial floristic pattern in even-aged stands of regenerated karri.

Strelein (1988) in his site classification of southern jarrah using floristics, similarly reported no relationships between floristic pattern and the level of disturbance either by fire or logging. These preliminary studies suggest that floristics are not affected to any great extent by felling and regeneration operations. Similar results have been reported by Purdie and Slatyer (1976) in eucalypt forests in south-eastern Australia and by others in different forest types (see Inions *et al.* 1990b).

Whether species composition and cover is related to age since disturbance (felling), is being examined in more detail in one karri community type (Wardell-Johnson personal communication).

Exotic weed species do establish on recently logged areas, and may increase in numbers if the interval between logging and regeneration is delayed for any reason. This is one reason why planting is used extensively instead of relying solely on the often irregular karri seed supply for regeneration (Breidahl and Hewett 1992). The weed species decline as regeneration develops.

Species abundance and cover values are known to vary significantly with time since fire (Christensen and Kimber 1975), which is part of the felling operation. Biomass and biomass accumulation (Grove and Malajczuk 1985), forest density (Stoneman *et al.* 1988) and forest structure all vary with time since disturbance. All of these, except forest structure, recover quickly.

Disturbance to animal communities following logging is also temporary. It is known that small ground mammals such as the southern bush rat (*Rattus fuscipes*), the mardo (*Antechinus flavipes*) and *Sminthopsis griseoventer* (Christensen unpublished), as well as larger animals such as the western grey kangaroo, the quokka, the southern brown bandicoot, the brushtail possum and the western ringtail possum all return to karri regrowth areas within two to 10 years following regeneration (Christensen *et al.* 1985; and Christensen unpublished). The mardo was recorded in very high numbers in one young karri regrowth stand where they were often located in bird-nesting boxes provided for hole-nesting birds (Wardell-Johnson 1986).

Tingay and Tingay (1984) found that clearfelled sites in karri forest initially attract small numbers of a few species of birds which are absent or uncommon on older sites. Young regeneration is colonized by bird species which prefer low and dense vegetation, and populations of these are larger than in mature forest. Wardell-Johnson (personal communication) has recorded similar results in an ongoing and more detailed study using mist nets.

Tingay and Tingay (1984) also found that most species which occur in mature forest were found to be present in 50-year-old regeneration. However, abundance of some species fluctuated between seasons compared with areas where these species were found in mature forest. Some species found in mature forest were absent or uncommon in young forest.

Reptiles, including the tiger snake (*Notechis ater occidentalis*), the mourning skink (*Egernia luctuosa*), Smith's skink (*Egernia napoleonis*), and the burrowing skink (*Hemiergis peronii peronii*), are all known to inhabit regrowth karri forest within a few years of regeneration. Frogs including *Geocrinia rosea* and *Crinia georgiana*, have been captured in pit traps within one year of regeneration in karri near Pemberton (Liddelow[15] personal communication). Wardell-Johnson (personal communication) is currently examining the long term effects of thinning and logging on birds and terrestrial vertebrates in one karri forest community.

Growns (personal communication) detected changes in the species composition of aquatic invertebrates in karri forest streams during and immediately following clearfelling and regeneration. This work is being continued into the regeneration phase as the regrowth forest develops.

Curry *et al.* (1985) examined short-term effects of clearfelling, burning and regeneration in karri on arachnid communities. They found no prolonged reduction in species richness but recorded changes in composition and structure of the arachnid community immediately following clearfelling.

No studies have been done on the recovery of karri forest arthropod communities following felling and regeneration. It seems likely, however, that there are similarities with the effects of fire on arthropods. In a review of the effects of fire in south-west forest and woodland ecosystems, Christensen and Abbott (1989) conclude that there is

> an immediate decrease in total numbers following a fire ....on the matter of recovery there is considerable conjecture, but it can be concluded that rate of recovery varies between different groups.

Obviously with clearfelling, the recovery would be expected to take longer since forest structure in particular takes a considerable time to recover.

Selection felling as practised in the jarrah forest appears to have minimal effect on flora and fauna. During the course of the many biological surveys carried out in the southern forests during the 1970s and early 1980s (Christensen *et al.* 1985), no noticeable deleterious effects of selection cutting were observed. Indeed, some species were noticed to be more common in selection cut areas, e.g. the western ringtail possum was most commonly observed in the Perup forest by spotlight at night in regrowth jarrah trees. It is also notable that the Perup Nature Reserve, one of the richest mammal fauna areas in the south-west, has a long history of selection

---

[15] G. Liddelow, Department of Conservation and Land Management, W.A.

felling. It has also been suggested that the numbat may be encouraged in certain areas by light felling which provides a greater range of hollow logs for refuges and dens (Christensen *et al.* 1984).

Hollow-nesting species form a group of particular concern with respect to clearfelling. In the karri forest, nine species of mammals (30 per cent of the total) require hollows in trees, (Wardell-Johnson and Christensen unpublished report). Wardell-Johnson (1985) found that 14 of 44 spp. (30 per cent) of birds observed in his karri forest study site used hollows in trees as nest sites. In a study of the use of nest boxes in 14-year-old karri regeneration, Wardell-Johnson (1986) suggests that hollows may be limiting for mardos in regrowth forest. Work is continuing on the effect of felling on tree hollow-nesting species.

No detailed research on the use of hollows has been carried out in the karri forest. A study that has just been completed in the northern jarrah forest showed very low occupancy of available hollows (Faunt[16] personal communication).

In summary, felling and associated regeneration burning create a temporary disturbance to the understorey layers, and longer lasting changes in the upper tree strata. On clearfelled areas where fire is used in the regeneration, plants recover quickly in a manner similar to that observed following a wildfire. The relative abundance and cover of species may change for a period, but these soon return to pre-disturbance values.

Forest structure takes longer to recover, and while a new canopy layer is soon formed, the trunks of the trees and branches take a long time, perhaps more than 100 years in karri, to develop quantities of holes suitable for hollow-nesting species. In karri forest where clearfelling is practised, the network of reserves and uncut zones along rivers, streams and roads, together with uncut swamp edges and the surrounds of granitic monadnocks, provide mature trees with hollows for animals and birds recolonizing the regenerated areas. Further research is needed to fully understand the role of this network of mature forest, and to determine the optimum distribution of uncut reserves and zones. The question of additional habitat in the form of retained mature trees within clearfelled areas is also being investigated.

The return of terrestrial animals, including invertebrates, appears to follow a pattern similar to that after a fire. However, there may be groups of invertebrates, such as relict groups of spiders and others with poor powers of dispersal, which may have difficulty reinvading regenerated areas, as postulated by Main B.Y. (1987). Such species, however, are mostly not located in those parts of the karri forest where clearfelling is practised.

Most plant and animal species appear to be able to survive disturbance caused by felling and fire. Further research should concentrate on species considered to be vulnerable: plants which have restricted distribution and are obligate seeders, and animals which depend on hollows in trees for nesting, or are relict species with restricted distribution and which may be sensitive to fire.

Recreation and Landscape

Forestry has a tradition of accommodating recreation. In the karri forest area in the years prior to World War II, activity was seldom planned or actively promoted, but when it occurred spontaneously, or was promoted by outside bodies, the Forests Department always welcomed interest and cooperated with acceptable initiatives. In the karri forest, the Rainbow Trail, a scenic drive which used Forests Department tracks, became one of the most popular tourist routes near Pemberton. The Gloucester Tree, a tree-top fire tower was also popular. Places of special and/ or historic value such as the Cascades, One Tree Bridge and the Four Aces received recognition and management attention.

[16] K. Faunt, Department of Conservation and Environment, Vic.

There has always been public interest in the activities of the timber harvesting industry. People are interested in forest management activity, viz., felling, hauling, regeneration, fire protection, fire towers and historical aspects of the timber industry which relate to the pioneering spirit of the early settlers. State forests provide a wide range of recreational opportunities.

With the gradual introduction of multiple use in the 1970s, the Forests Department took a more proactive role in recreation management. Areas already established as popular venues, e.g. Gloucester Tree, Mt. Frankland, Cascades and One Tree Bridge were declared MPAs for recreation and were protected by areas of forest designated for special silvicultural treatment. Other areas considered to have recreational potential were protected by MPA zoning in advance of public 'discovery', e.g. Muirillup Rock.

Mill and forestry settlements which closed down were retained and leased to recreational interests, e.g. Pimelea, Donnelly River, and Tone River. Scenic picnic spots along roads and at other special spots were developed; walking trails were established; drive-yourself scenic tours were devised; publications were produced. Recreation officers were appointed in the districts, and recreation values were taken into account in day-to-day operational direction.

Since the creation of CALM, MPAs for recreation have been given greater security and purpose by making them either national parks or conservation parks, both of which have recreation as a major recognized use. The Southern Region Management Plan (CALM 1987b) lists 53 recreation sites which may provide up to 15 different recreation opportunities: picnic barbecues, fishing, swimming, canoeing, marked walk trails, nature study interpretation of the natural environment, camping, lookout tree climbing, boat launching, scenic drives, photography and marked horse trails.

Of special interest are CALM's landscape planning skills which are now being applied to the karri forest. Computerized techniques are available which depict scenes from specified viewing points. These will allow landscape experts to plan road zones in an objective fashion. These skills are now utilized in most areas of forest management planning, for example designing cutting coupes, new roads, barbecue and picnic sites.

State forest provides a greater span of recreational opportunity than do national parks or nature reserves. Horses and dogs can be exercised, firewood can be collected, cubbies can be built, craft wood can be sought and purchased, and the rules covering fishing and camping are more lenient.

These, the forest operations themselves, and the roads essential for their execution, combine to provide an array of opportunities which essentially complement those of national parks. Recreation, tourism and conservation all profit from the active, healthy existence of the three categories of public lands, nature reserves, national parks and State forest. Much pressure from recreational pursuits is taken off national parks in particular, by catering for many tourist activities in State forests.

Protection

Protection covers all activities which defend the public forest estate from threats to its health or existence. Although this subject logically covers many aspects such as security of tenure and purpose, legislation, and law enforcement, in this context the main factor under consideration in the southern forests is wildfire, plant pests and diseases.

Those responsible for fire protection in the karri forest must live with four realities, viz.

- Ignition through human or natural (lightning) causes will eventually occur during periods of high fire danger.

- Wildfire cannot be controlled in fuels greater than 20 t ha^{-1} during periods of high fire danger.
- Fire is an integral factor of the karri forest ecosystem. All biological components have evolved adaptations to its occurrence.
- Wildfire cannot be accepted as an uncontrollable variable in forest management. The risks to life, townships, farms, plantations, regrowth destined for future timber production, tourism, recreation and nature conservation values are too high to gamble with the risk.

There is only one practical way in which the wildfire phenomenon in the southern forests can be brought down to one of human dimension: fuel reduction or fuel modification. Prescribed burning during mild weather is the simplest method of achieving fuel reduction. Prescribed burning does not, however, stop fires occurring, nor does it absolve forest managers from the effort and expense of being vigilant, organized, trained and well equipped. But with less fuel on the forest floor, fires can be more easily suppressed.

The history and development of prescribed burning, as well as fire effects, are discussed by Underwood and Christensen (1981). Documented evidence of the efficiency of the techniques in Western Australian forests is given in Underwood *et al.* (1985).

Current strategy involves periodic burning under mild conditions, the interval between burns depending upon the rate of fuel accumulation. In karri forest outside conservation reserves the burning rotation is usually about seven to nine years.

Fire use is not confined to a simple strategic protective role. In karri it is used at high intensity following timber harvesting to promote regeneration. Habitats for certain animals requiring fire-regenerated thickets are renewed by fires of appropriate intensity over limited areas (Christensen 1987b). Wildflower displays along roads may be encouraged by judicious use of fire. In some cases, fires need to be excluded from areas for long periods; karri regeneration needs to be kept free of fire for 15 to 20 years, while other areas are purposely kept unburnt as experimental controls (Christensen and Abbott 1989).

Strategically, CALM aims to keep wide east-west bands of the southern forests free of cutting and regeneration for extended periods. This allows managers to keep the fuels low on these areas, and so protect intervening zones within which regenerated forest is in the process of development. In turn, wide bands of similar aged regeneration are being established which can, after 15 to 20 years of age, become fuel reducible, and replace the uncut bands, which then become available for cutting and regeneration. This inevitably has a moderating effect on coupe dispersion and road location. The need for CALM to be able to plan decades ahead is evident.

Prescribed burning, in particular its effects on the environment, is a controversial subject. Considerable information is now available on the effects of fire, and current evidence suggests no reason why the practice should not continue in a well planned way which takes into account both the risks of fire, and the effects of burning (Christensen and Abbott 1989). The Australian Academy of Science (Anon. 1975) saw a role for fire in the management of ecological reserves in short

> fire is a key management tool in many areas, as well as the most controversial .... total protection from fire, while being impossible in the long term (except perhaps in rainforest situations), results in a population structure of the perennial plants, particularly woody plants, strongly skewed towards over-representation by mature individuals, with a strongly discontinuous age distribution, and an absence of many successive stages.....it is probable that a succession-based management strategy would involve a mixture of hot

and cold fires, each of restricted area and of a scale that only a small fraction of a reserve would be burnt in any one year, thus creating a mosaic of successive states and a consequent increase in habitat diversity.

Research studies have provided support for this view, and it is now well recognized that fire has a vital role in the management of conservation reserves (Christensen 1986, 1987a). Management plans for parks and reserves within the area, e.g. the Shannon-D'Entrecasteaux and Walpole-Nornalup National Parks management plans have burning plans that aim to introduce and maintain fire-induced diversity. Similar plans exist for the Denbarker area and the Lake Muir Nature Reserve System. Likewise, forest district management plans incorporate fire diversity in their prescribed burning. There is an extensive network of 'no planned burn' areas in place throughout the forest (Christensen and Abbott 1989) and species fire management plans are in place in the Perup Nature Reserve (Christensen and Maisey 1987).

Several pests are of economic importance in the forests of the south. In the early 1960s, populations of a native leafminer of jarrah, *Perthidia glyphopha*, erupted. By 1987 some 750 00 ha of forest were infested (Abbott 1990), much of it in the southern forests. Considerable economic losses were incurred through loss of growth every year (Abbott 1992). This pest is the subject of an ongoing research program aimed at understanding the underlying factors controlling outbreaks and searching for methods of control (see Abbott 1992). Some 20 per cent of trees show a high level of resistance to infestation. Current management techniques to minimize future growth losses include leaving all resistant trees regardless of size when thinning the jarrah forest.

A second native insect species, gumleaf skeletonizer (*Uraba lugens*), erupted in 1982, and by 1986 had infested about 240 000 ha of jarrah and marri forest. By 1991 no infestation was to be found (Abbott 1992), Strelein (1988) and Abbott (1990) have circumstantial evidence that the extent of the skeletonizer outbreaks since 1983 is greatest following dry, warm winters. Monitoring of the forest for this species is continuing.

Bullseye borer (*Trycopharia acanthocera*) causes damage to young karri and marri trees by tunnelling through the stems. It is also associated with brown-wood, a condition which precedes wood decay in karri (see Abbott in press).

Disease is also an important aspect of forest protection. The dieback disease fungus *Phytophthora cinnamomi* occurs in many of the communities in the southern forests. Most species in the karri forest are not prone to *Phytophthora cinnamomi* attack, but a wide range of the flora in the non-karri vegetation associations are susceptible, in particular, members of the families Proteaceae, Epicridaceae and Mrytaceae. Detailed planning and operational procedures exist to restrict the spread of dieback disease and these are observed in all forest operations, such as forest harvesting and regeneration, road building and maintenance and recreation (Underwood and Murch 1983; Shearer and Tippett 1989).

Harvesting and Silvicultural Operations

The extent to which various aspects of forest management previously discussed influence conservation in the forest is largely determined by the rate and the manner in which trees are cut from the forest. Firstly, to continue to sustain timber yield from the forest the 'allowable cut' must be determined, and secondly, there must be sufficient controls to ensure that the cut is achieved in the correct manner.

History, politics, social factors, as well as science, all play a part in forest management and the achievement of economic and conservation goals.

The way logging operations have been carried out has changed over the years. Before 1987, the sawmilling company was responsible for all logging operations. At that time CALM's (previously the Forests Department's) main tool to enforce its direction of bush operations was the issue of a licence to forest workers, without which a person was not legally entitled to be employed in State forest. Threats to suspend, and sometimes actual suspension, of the licence were used from time to time to correct malpractice. In addition, the annual renewal of rolling five-year logging and regeneration plans allowed CALM to control the location and timing of logging activity in the forest.

During the 1980s the increasing complexity of harvest operations, with integrated operations supplying up to 15 different types of product, saw a new system of integrated logging develop. CALM now takes direct responsibility for harvesting operations. In effect, contractors now work for CALM rather than the sawmilling companies.

In the 1970s and 1980s sophisticated new methods were implemented to assist in the planning, control and recording of harvesting and regeneration operations. This revolution in procedure was made necessary by the introduction of integrated operations, the requirements of multiple use management, and techniques to minimize the spread of dieback disease. The procedures have been made possible by the computerization of inventory data, planning data, and harvesting data records and environmental information, the use of aerial photography and the design of a special Hardwood Operations Control System (HOCS).

Planning, including fire and disease protection, silviculture, road making, long-term yields, coupe design and anticipated wood product demand, now takes place at two levels: strategic planning (four years ahead) and tactical planning (one year ahead).

First stage planning (strategic, four years ahead)

Land Use	-	buffer zoning; dieback disease quarantine
Fire protection	-	location of strategic buffers
Dieback	-	mapping; hygiene and impact analysis; needs to be three years ahead of operations
Silviculture	-	karri seed forecasts; nursery planning
Product mix	-	anticipated demands for various log products
Roading	-	alignment selection and construction well ahead of operations
Long-term yields	-	inventory at management level
Coupe design	-	size; dispersion; product mix required
Visual resource	-	visual quality zones
Special care zones	-	identify and demarcate

Second stage planning (tactical, one year ahead)

Coupe	-	boundaries; forest types; sensitive areas establishment (slopes etc)
Silvicultural prescriptions	-	tree marking
Fallers blocks	-	demarcation
Environmental management		
	-	seven-way tests for dieback disease
	-	plans for phased logging if required
	-	on-the-ground demarcation of road, river, and stream zones

	-	on-the-ground demarcation of special care areas
Fire protection	-	prescribed burning of uncut surround
	-	advanced burning if required
Aerial photos	-	for HOCS and planning use

During logging

Supervision of Silvicultural Prescription

- damage to retained stems
- monitoring soil disturbance
- installation of erosion control measures

Classification of product mix (the various wood products which come out of a coupe)

Inspections to certify satisfactory completion of fellers blocks, landings and coupes

Recording of progress in HOCS

Post-logging

Regeneration operations	-	coupe preparation
	-	regeneration burning
	-	regeneration assessment
	-	planting and seeding as specified
	-	rehabilitation of disturbed soils
Recording of progress in HOCS		
	-	aerial photographs
Fire protection	-	indefinite

The increase in the number of functions caused by the introduction of multiple use management has also changed the traditional pyramid type line of command within an area. Today, operational direction is given by a group, consisting of representatives of the various functions, e.g. timber production, fire and environmental protection, recreation, conservation and research.

Summary

In Chapters 2 and 4 it was demonstrated that the southern forests, which include the forests used for timber production, are an extremely valuable conservation asset. The conservation reserve system and the multiple use forest area cannot be considered in isolation; they occur in a mosaic in which the boundary of one grades intimately into that of the other. The management of one therefore inevitably affects the other.

Almost all aspects of forest management for timber production include environmental guidelines and checks and balances to ensure that forest operations are carried out in the most environmentally sensitive manner. Continuing research ensures that there is continued updating and upgrading of management practices. Many extraneous factors, social and political, impinge on the system to put pressures on management which may affect conservation values of the forests. The scientifically 'best' way is not always possible; often forest management decisions are made by 'trade offs' between values which society demands from the forests.

State forests are a valuable conservation asset which are managed in sympathy with the conservation reserve system, immeasurably increasing the value of the latter.

In this respect, forest reserves differ fundamentally from other reserve systems in the south-west and indeed much of the rest of Western Australia. The limited value of the wheatbelt nature reserve system, small reserves surrounded by biologically inhospitable wheatfields, is well recognized (Kitchener *et al.* 1980; Kitchener 1982).

CHAPTER 7

EVALUATION - ADEQUACY OF CONSERVATION WITHIN THE KARRI AND SOUTHERN FORESTS

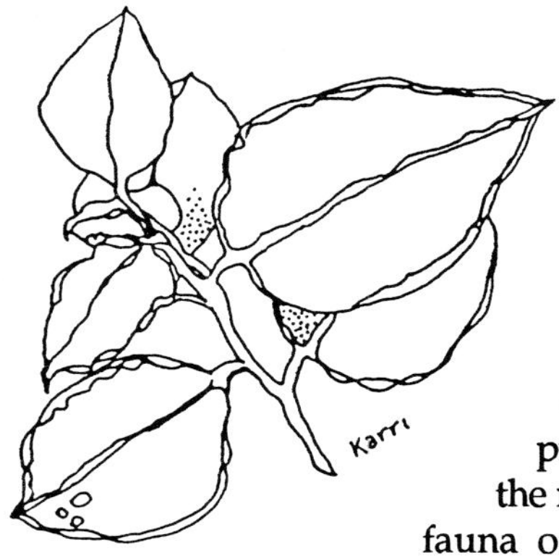

During the Paleozoic and Mesozoic eras, Australia had a rich and diversified biota which it shared with Pangaea, or the rest of Gondwana. Following the breakup of Gondwana, the climatic changes which occurred during the Tertiary, together with the northward movement of the Australian continent, resulted in marked changes in the biota. The predominantly subtropical rain forest vegetation changed to the more dry climate-adapted plant associations and associated fauna of today. In south-west Western Australia xeromorphic communities, which were a minor component of the early Tertiary flora, proliferated and expanded so that today only minor components of the original flora remain.

The southern forests are the southernmost and the wettest portion of the south-west. They are unique in that they contain:

- All of the tingle (*E. guilfoylei, E. jacksonii* and *E. brevistylis*) forest formations in existence, and almost all of the karri forest.
- A portion of the highly speciated, autochthonous xeromorphic heathland flora. A very high percentage of species are endemic, if not to the southern forests, to the south-west.
- A 'mesic' element of the sclerophyllous heathland flora unique to the southern forests that may be directly descended from the ancient xeromorphic element of the subtropical Tertiary flora. It is speculated that these mesic sclerophyll heathland communities may be most like the original xeromorphic flora which grew under high rainfall conditions in the Tertiary. The more extensive 'dry' sclerophyllous heath communities of the medium rainfall zone, although descended from the same Tertiary xeromorphic communities, appear to have speciated and changed to a far greater extent than their mesic counterparts.
- A mesic fauna with a high proportion of relict forms among the lower animals, the frogs, fishes and the invertebrates, especially the aquatic invertebrates.
- A range of rare and restricted species which survive nowhere else.

The high precipitation, which over millions of years has contributed to the evolution and survival of the unique flora and fauna, has more recently been a major indirect threat to their survival. Early European settlers were first attracted to the area primarily because of the high rainfall and equable climate. On the face of it this would appear to lead to direct conflict between conservation and development as is so frequently the case. Fortunately, other factors have largely prevented this, foremost among these are the area's soils and its forests.

The nutrient-poor soils, which since the early Tertiary have featured so largely in the development and maintenance of the unique sclerophyllous vegetation, continue to protect their biological offspring. These poor soils are the very worst for agriculture, and as a consequence, most remain uncleared. Yarri forests on the rich red loams, and wandoo forest and woodland and jarrah on the better soils have borne the brunt of clearing for agricultural. These better quality soils generally support plant communities with comparatively low species diversity. Another feature in favour of the forest is low mineralization. Unlike the northern jarrah forest, the karri forest has been of little interest to the mining industry.

Forests have contributed to the conservation of the area's flora and fauna because extensive tracts were protected from clearing for agriculture because of their reservation for the production of timber. Timber production forests, particularly when managed sympathetically, have considerable conservation value: few species of vertebrates have disappeared from Australian forests over the last century (Recher 1986; Recher *et al* 1987). In the southern forests, forestry activities have also had a minimal impact on flora and fauna (Christensen *et al.* 1985; Hopper *et al.* 1992). For these reasons the total proportion and area of land in the southern forests with a conservation status, i.e. national and conservation parks, nature reserves and State forest, is high.

The Level of Reservation

Flora

A breakdown of land tenure in the southern forests in 1991 is given in Table 4. Of a total of some 2 114 700 ha, 1 322 400 ha (62.5 per cent) is in reserves, parks or State forest. Of this total, some 381 800 (28.9 per cent) is in national parks, nature reserves and other conservation reserves; 940 600 ha (71.1 per cent) is dedicated to multiple use forestry. A further 76 100 ha of this area is road river and stream zones, and 143 100 ha is heathland, sedgeland, woodland and rock outcrops reserves, managed primarily for conservation purposes.

Table 4

LAND TENURE IN THE SOUTHERN FORESTS AS AT 31 July 1991

	FORESTED AREAS	WOODLANDS HEATH/SEDGELANDS & OTHER	LAKES & WATER	SUB-TOTAL (HA)
CALM Land *				
National Park	129 400	149 100	1 100	279 600
Nature Reserve	68 900	12 700	7 700	89 300
Conservation Park	12 800	100		12 900
Multiple use forest (Production)	796 800	143 100	700	940 600
Total CALM Land				**1 322 400**
Other Government Land (Owned or vested)				**58 200**
Private Property				**734 100**
TOTAL				**2 114 700**

* As proposed by CALM (1987b).

N.B. There is a small area on the south coast which is not included in these figures because the total area is not readily available. However, the area of CALM land, which is available, totals a further 18 114 ha, 17 585 of which is national parks and nature reserves.

Probably no part of the southern forests could be considered to be completely 'natural', i.e. free of any human impact (Scott 1970). Most State forest has been subject to some disturbance by humans or fire. However, disturbance is essential to the maintenance and regeneration of certain ecosystems, (Mount 1964; Webb 1970; Heinselman 1971) and some fauna (Davey 1990). The communities and ecosystems contained within State forests can therefore be expected to contribute significantly to conservation in the area.

The value of a reserve system is not to be measured by quantity alone. Quality and distribution are also important factors. Kitchener *et al.* (1980) and Kitchener (1982), working in isolated wheatbelt reserves, demonstrate convincingly that it is the number of different habitats contained in a reserve rather than size *per se* which dictates species richness. Similarly, Burgman (1988), working in the semi-arid sandplain and mallee near Esperance, has demonstrated that geographic distance is important in predicting the resemblance between stands, both between and within vegetation formations. Thus, it is not sufficient to allocate reserves to a representative stand in each vegetation formation without accounting for the spatial scale of floristic pattern.

This leads to the question of what exactly should be conserved. Ideally, it is desirable to conserve the full spectrum of ecosystems, as well as the full range of plants and animals characteristic of the area. This is rarely possible. The most appropriate approach appears to be to conserve as full as possible a range of ecological types characteristic of the area. This is the approach advocated by Specht and Cleland (1961) in South Australia, Webb (1966) in Queensland and Specht *et al.* (1974) for Australia and New Guinea.

While there is probably little disagreement on this principle, the level at which selection and definition of areas should and indeed can take place will continue to be a matter of controversy. As pointed out by Austin and Miller (1978) and Heddle *et al.* (1980), the structural approach used by Specht *et al.* (1974) is limited in its discriminatory power. Floristic classifications such as those of Havel (1975a) offer a more precise classification with a greater level of discrimination, but still have their limitations.

Three questions arise:

- what classifications are available for the plant communities of the area;
- what level of discrimination is possible; or
- what level of discrimination is sufficient and adequate?

In assessing the reserve system for the northern jarrah forest, Havel (1989) was able to examine the conservation adequacy quantitatively at the level of vegetation complexes, that being the highest level of complexity at which the area is mapped. At the more precise level of site vegetation types (see Havel 1975a, 1975b), he was only able to examine the situation in qualitative terms by relating site-vegetation types to vegetation complexes.

In the southern forests, the most complete classification of the vegetation for which maps exist, is that of Smith (1972, 1973) who uses the structural systems of Specht (1970) to identify major formations together with species composition to association level. The area to the east of the Frankland River, not covered by Smith, was mapped by Beard (1980). In both cases, mapping is to the level of formation and major associations only, and not detailed enough for quantitative determination of plant associations, so assessment has to be semi-quantitative. That is, it is only possible to list the plant associations contained in each reserve, not the percentage or the area which is covered by each of them.

Accordingly, plant associations have been listed by reserves where they are known to have a significant representation (Appendix 6). A total of 51 plant associations in 19 formations recognized by Smith (1972, 1973) are included. As well, further associations from Beard (1980) and 20 'new' associations recognized by the author and A.R. Annels are listed, forming a combined total of 82 native and one introduced plant associations.

All of the recognized plant associations are represented in one or more of the parks and reserves of the southern forests (see Appendix 6). A few are poorly represented in reserves and these are listed below.

- High open forest and open forest of yarri: there is very little of this forest type left, most having been alienated and cleared for agriculture before World War II. There may be small areas worthy of reservation still to be found in multiple use forest areas.
- Low open forest and low woodland of jarrah: there is little of this type in the reserve system, but large areas exist on Crown land in the Kent River catchment which could be considered for reservation.
- Scrubland: an association characterized by *Pultenaea reticulata* and other species which occur on quartzite sands in certain upland swamps. This association occurs especially in the Donnelly River system. Much of this type is protected in the heathland, sedgeland, woodland and rock outcrop reserves.
- Sedgeland on the Margaret River: these associations are poorly represented in the reserve system, but are not 'threatened' by land use or management.
- Samphire heath: there are only limited areas of this type within the area.
- *Kingia australis* high open scrub: extensive areas of this association exist in the Donnybrook Sunklands just to the north of the Blackwood River. Some of this area could be considered for reservation.

The plant communities listed are described more fully by Smith (1972, 1973), with the exception of scrubland which is recognized by the author and Annels.

Further definition to the level of the site vegetation types of Strelein (1988), McCutcheon (1978 and 1980) and Christensen (1980b) for the jarrah forest formations and Inions *et al.* (1990a, 1990b), and Wardell-Johnson *et al* (1989) for the karri is possible at a qualitative level. It is highly unlikely that quantitative determination to the level of site-vegetation types will be available for a very long time. Mapping at this level can be done effectively only by ground traverse, which is exceedingly difficult and expensive. It is unlikely that it will be worth doing for a long time to come.

However, all site-vegetation types recognized in these studies are known to be represented in the reserve system. Two of the karri types of Inions *et al.* (1990b) are believed to be restricted entirely to national parks.

Having examined the representation of plant associations and communities, it is pertinent to ask: are all areas of special scientific significance which are known in the area well represented in the reserve systems? A total of seven such areas are identified in this review. These areas are listed below together with brief comments on their conservation status and their scientific value. More detailed information on the scientific value of each area is to be found in Chapters 2 and 4.

- *The tingle forest- yellow tingle (Eucalyptus guilfoylei), red tingle (E. jacksoni) and Rates's tingle (E. brevistylis)*

This forest contains a relictual mesic fauna, particularly among the invertebrates. Almost all of the area of red and yellow tingle forest is within parks and reserves. Rate's tingle is a minor component of other plant associations which are well represented in reserves.

- *Granitic monadnocks*

These may contain geographic relicts which are often of Gondwanan origin. Many of the more prominent occur within parks and reserves; most of the remainder are in State forest where they are secure because of their inaccessibility and the absence of any commercial value.

- *The southern sedgelands and wet sclerophyllous heaths*

These contain relictual forms of Gondwanan and pre-Gondwanan origin among the aquatic fauna. They are also a centre of speciation for a range of Monocotyledonous families. Of all of the present day plant associations in the south-west these may well be the closest to the original xeromorphic flora of the early Tertiary. These are very well represented in the Walpole-Nornalup and Shannon-D'Entrecasteaux National Parks. The remainder are in State forest where they are subject to minimum disturbance, and form part of the heathland, sedgeland, woodland and rock outcrop reserve system.

- *The Perup Nature Reserve*

This contains six species of endangered mammals. These are especially important as they all belong to the group characterized by Burbidge and McKenzie (1989) as being in the 'critical weight range' (CWR) group, a section of the mammal fauna which has fared very badly since European colonization.

The entire area of 40 200 ha, a former Forests Department Fauna Management Priority Area, is proposed as a nature reserve.

- *Parts of the Donnybrook Sunklands, especially the Wicher Range*

Contains rare and restricted flora and geographical outliers.

These are well represented in reserves, with the exception of the *Kingia australis* high open scrubland associations and the Margaret River sedgelands. However, neither of these vegetation types are likely to be affected by timber cutting.

- *The Leeuwin-Naturaliste Ridge and much of the south-west coastline from Cape Naturaliste to Torbay*

This extensive area contains geographic and Gondwanan relicts, particularly among the invertebrate fauna, rare species of plants and archaeological sites of significance.

Almost the entire coastline is within national park, or proposed national park.

- *The Mt. Lindesay - Sheepwash Creek area*

The junction of three major botanical districts, the Warren, Stirling and Darling of Gardner and Bennetts (1956). The area contains representative plants from each of these major botanical regions, and as a consequence has a very rich flora. A considerable portion of the area is within a proposed national park (CALM 1987b). Further additions as recommended by Christensen (unpublished report 1980a) could be considered.

Two further considerations are of importance with regard to the adequacy of the reserve system. The first is well put by Burgman (1988) who states

> there is much to be said for a reserve system that includes most species regardless of any other information. However, this approach assumes that the probability of survival of a species is dependent only on its inclusion in a reserve. It ignores the spatial distribution of land allocation and the relative abundance of different species. A better solution would be a reserve system that includes at least all rare plants. Best of all, a reserve system should include all plants likely to face extinction as a result of the proposed land use practices, and include them in sufficient numbers to ensure some acceptable risk of extinction.

The important points embodied in this quote are firstly, the assumption that the probability of the survival of a species is dependant on its inclusion in a reserve. The multiple use State forests, which surround the majority of the reserves in the southern forests, provide a high degree of security for populations of species.

The second point is that all plants likely to face extinction should be included in reserves. The southern forests are comparatively low in numbers of species listed in rare and restricted categories (see Table 2b), and a large proportion of plant species which are listed as declared rare flora are represented in the reserve system (see Table 1). Many of the rarer plants, monitored species and species of uncertain status, on the other hand, are not within reserves but are found within State forest or are on areas of private property (see Table 2a). Since further undescribed species may still be discovered as the flora of the area is more thoroughly surveyed, this is an aspect which will require constant review.

Genetic variation in karri

The question of conservation of the genetic resource of karri itself has received scientific attention. Coates and Sokolowski (1989) studied the level and geographic distribution of genetic variation of karri at 16 polymorphic allozyme loci. They found that genetic variation was higher within than between separate populations. The outlying karri populations (e.g. in the Porongurup Range and at Mt. Manypeaks) exhibited noticeable levels of genetic divergence from the main karri forest belt, but no significant divergence was found between populations in the main karri forest.

All of the major allozyme variants for karri are contained within conservation reserves except for a small population at Rocky Gully which is totally within private property (Hopper and Coates 1990).

Fauna

A major difference between flora and fauna conservation is the level at which conservation is attempted, i.e. at the community or the species level. Vertebrate species richness is closely linked with habitat factors such as the number and variety of plant communities present (Kitchener *et al*. 1980; Kitchener 1982). Conservation of flora at the community and site-vegetation type level therefore caters to some extent for fauna, particularly the invertebrates (see Worsley Alumina 1985), which are as yet incompletely known. For the much smaller number of vertebrate species, evaluation of the adequacy of their conservation is possible at the species level.

Mammals

The status of the mammals was discussed in Chapter 1 and is indicated in Table 5. The six extant species endemic to the south-west which occur in the southern forests, the honey possum (*Tarsipes rostratus*), the quokka (*Setonix brachyurus*), the western brush wallaby (*Macropus irma*), the western ringtail possum (*Pseudocheirus occidentalis*), the grey-bellied dunnart, *(Sminthopsis grisoventer)* and *Fallistrellus mackenziei* are each well represented in several reserves and parks within the area. There is concern, however, for the western brush wallaby which is still common but appears to be declining throughout its range (Christensen unpublished data), owing possibly to fox predation.

The six species which are declared to be threatened under the Wildlife Conservation Act (1950-1979), the woylie[17] (*Bettongia pencillata*), the tammar wallaby[17] (*Macropus eugenii*), the chuditch[17] (*Dasyurus geoffroii*), the numbat[17] (*Myrmecobius fasciatus*), the western ringtail possum and the southern brown bandicoot (*Isoodon obeselus*) are well catered for by the reserve system. The major area of occurrence of all but the last two of these species within the area is in the Perup Nature Reserve. This reserve contains the main populations of the woylie and the largest mainland populations of the tammar wallaby. The southern brown bandicoot is common in suitable dense habitat in many areas of the southern forests. Good populations of the western ringtail exist in the Perup Nature Reserve and the Leeuwin-Naturaliste National Park.

[17] There are active management programs to increase the numbers and distribution of these species.

Table 5

STATUS OF MAMMALS OF THE SOUTHERN FORESTS

Species	Current Status
Western grey kangaroo *(Macropus fuliginosus)*	S (stable)
Western brush wallaby *(Macropus irma)*	D (declining)
Tammar wallaby *(Macropus eugenii)*	S
Quokka *(Setonix brachyurus)*	S
Woylie *(Bettongia penicillata)*	S
Burrowing bettong *(Bettongia lesueur)*	E (extinct)
# Broadfaced potoroo *(Potorous Platyops)*	E
Gilberts potoroo *(Potorous tridactylus gilbertii)*	E
Common brushtail possum *(Trichosurus vulpecula)*	S
Western pygmy possum *(Cercartetus concinnus)*	S
Western ringtail possum *(Pseudocheirus peregrinus occidentalis)*	D
Honey possum *(Tarsipes rostratus)*	S
Southern brown bandicoot *(Isoodon obeselus)*	S
Bilby or dalgyte *(Macrotis lagotis)*	E
Western quoll or chuditch *(Dasyurus geoffroii)*	S
Brush-tailed phascogale *(Phascogale tapoatafa)*	S
# Red-tailed phascogale *(Phascogale calura)*	E

Table 5 *(continued)*

SPECIES	CURRENT STATUS
Yellow-footed antechinus or mardo *(Antechinus flavipes)*	S
Grey-bellied dunnart *(Sminthopsis griseoventer)*	S
Numbat *(Myrmecobius fasciatus)*	D
Bush rat *(Rattus fuscipes)*	S
Water rat *(Hydromys chrysogaster)*	S
Gould's long-eared bat *(Nyctophilus gouldi)*	S
Greater long-eared bat *(Nyctophilus major)*	S
Lesser long-eared bat *(Nyctophilus geoffroi)*	S
Gould's wattled bat *(Chalinolobus gouldii)*	S
Chocolate wattled bat *(Chalinolobus morio)*	S
King River Eptesicus *(Eptesicus regulus)*	S
White-striped mastiff bat *(Tadarida australis)*	S
Little mastiff bat *(Mormopterus planiceps)*	S
(Falsistrellus mackenziei)	S
Echidna *(Tachyglossus aculeatus)*	S

* 3 extinct, possibly 5
3, declining
24, stable

\# Uncertain whether they occurred within the area.

All the other species of mammals which occur within the area (Appendix 3a) are well represented in the conservation reserve system. Most of them also occur widely in State forest (Christensen *et al.* 1985). The conservation status of the numbat, the western brush wallaby and the western ringtail deserves attention. An attempt to breed the numbat in captivity is currently underway and a reintroduction program is in progress. Further studies on the biology and the protection of the other two species are needed, together with improved programs of predator (fox) control.

Birds

The great variety of the flora and of vegetation communities in the southern forests contributes to the comparatively rich bird life of the area. Most species of birds are relatively common and most occur in each of several major plant formations and communities (Christensen *et al.* 1985) which are all well represented in the reserve system. Birds also cope fairly well with, and are sometimes favoured by, silvicultural operations (Recher *et al.* 1987) and consequently continue to be common in multiple use forest areas as well as in national parks and nature reserves. All significant wetlands on public lands within the area are reserved providing security for waterbirds. Many waterbirds have also benefited from the numerous dams and reservoirs within the area.

Plate 1 High open karri (*Eucalyptus diversicolor*) forest on the South West Highway (photo C.Winfield)

Plate 2 High open tingle (*Eucalyptus jacksonii*) forest (photo G.Liddelow)

late 3 Open jarrah (*Eucalyptus marginata*) forest photo T.Leftwich)

Plate 4 W.A. peppermint (*Agonis flexuosa*) low open forest (photo G.Liddelow)

Plate 5 Wandoo woodland (*Eucalyptus wandoo*) (photo T.Leftwich)

Plate 6 Melaleuca (*Melaleuca cuticularis*)low woodland with *Hibbertia stellaris* (photo T.Leftwich)

Plate 7 Sedgelands frequently occur in mosaic with forest to the south of the main karri belt (photo T.Leftwich)

Plate 8 Mt. Chudalup, a typical granite monadnock fringed with karri and jarrah forest merging to woodland and closed heath (photo G.Liddelow)

Plate 9 Lake fringed with paperbark (*Melaleuca preissiana*) woodland (photo T.Leftwich)

Plate 10 Pool on Frankland River (photo G.Liddelow)

Plate 11 Point D'Entrecasteaux, the feature from which the National Park takes its name (photo C.Winfield)

Plate 12 Tammar (*Macropus eugenii*) wallaby - The largest mainland population of this species occurs in the Perup Nature Reserve (photo T.Leftwich)

Plate 13 The western ringtail possum (*Pseudocheirus occidentalis*). The major remaining populations of this species occurs within the southern forests (photo T.Leftwich)

Plate 14 The numbat (*Myrmecobius fasciatus*). This species is now restricted to the south-west of Western Australia where a significant population occurs in the Perup Nature Reserve (photo T.Leftwich)

Plate 15 The short-nosed snake (*Notechis minor*) a species largely confined to the southern forests (photo J.Wilson)

Plate 16 *Geocrinia lutea* a small frog with a very restricted distribution confined to the southern forests (photo G.Wardell-Johnson)

Plate 17 *Lepidogalaxias salamandroides* the sole member of the family Lepidogalaxiidae a small fish of very ancient lineage which is almost entirely restricted to the southern forests (photo G.Liddelow)

Plate 18 *Moggridgea tingle*, a small Gondwanan trapdoor spider with African affinities from the tingle forests near Walpole (photo B.Y.Main)

Plate 19 *Anthocercis sp.* a rare shrub believed to be restricted to the Walpole-Nornalup National Park (photo G.Wardell-Johnson)

Plate 20 Regrowth karri resulting from regeneration following clearfelling near Boranup in the 1880s (photo T.Lefwitch)

Plate 21 Regrowth karri, recently thinned, resulting from regeneration following clearfelling near Big Brook Pemberton in the late 1920s (photo T.Leftwich)

The one species on the declared list, the crested shrike-tit (*Falcunculus frontatus*) continues to maintain its numbers in the area (Storr 1991). Never common, it is nevertheless widespread in the area and well represented in the reserve system.

Suitable coastal habitat exists (most of it within national parks and nature reserves) for five of our rarest birds, the western subspecies of the ground parrot (*Pezoporus wallicus flaviventris*), the noisy scrub-bird[18] (*Atrichornis clamosus*), the western bristle-bird (*Dasyornis longirostris*), the western subspecies of the Rufous bristle-bird (*Dasyornis broadbenti littoralis*) and the western whipbird (*Psophodes nigrogularis*). All occurred within the area in historic times, mainly in coastal heathlands, but none have been sighted for many decades.

Reptiles, amphibia and fish

The majority of the known species in these groups are relatively common and widespread in the area (Christensen *et al.* 1985), and are well represented in national parks and nature reserves. Two rare snakes, the short-nosed snake (*Notechis minor*) and the square-nosed snake (*Rhinoplocephalus bicolor*) are uncommon, but a major portion of both their ranges lies within the coastal parks and reserves (Christensen *et al.* 1985).

The most interesting frogs, the *Geocrinia rosea* complex, have extraordinarily limited distributions. *Geocrinia rosea*, which occurs in the Donnelly, Warren and Shannon River valleys, is well represented in the reserve system, both in parks and in the river and stream zones. *Geocrinia lutea* is largely confined to swamps in the Walpole-Nornalup National Park.

Two recently discovered species which are now included on the gazetted list of rare fauna, *Geocrinia vitellina* and *G. alba*, are vulnerable because of their limited range and unusual breeding biology (Wardell-Johnson and Roberts 1991a). *G. alba* is confined to a few sites, many of them uncleared land on private property, in the Karridale area. A portion of its range is within the Leeuwin-Naturaliste National Park. Protection of suitable creek habitat will be essential to ensure the survival of this species. *G. vitellina*, possibly Australia's most restricted vertebrate, is confined to a few seepage sites in an area of less than 0.2 km^2. This area is within State forest, but not within the reserve system. A special management plan to protect the frog's habitat is being prepared (Wardell-Johnson and Roberts 1991b).

The various small fish which are endemic to the south-west are all widespread in the area and are locally common. All are present in the reserve system (Christensen 1982). Recent studies also suggest that logging operations have a minimal effect on water quality where stream buffers are present (Borg *et al.* 1987b). The area of forest with river and stream zones therefore adds considerably to the conservation status of these small fish.

Invertebrates

Several groups of invertebrates from the area, many of them endemic with Gondwanan affinities, are currently attracting considerable attention. New species and genera are turning up among the aquatic invertebrates, including the larger crustaceans. It is likely that many of these will be endemics, some of them of ancient lineage.

Among the spiders, new species of relict forms are also turning up from the moister habitats in the south of the area, particularly in the tingle forest. Consideration will need to be given to the conservation of such species when more information comes to hand.

---

[18] The noisy scrub-bird is being reintroduced to the area at the present time.

Other values

Old-growth forest

'Old-growth' forest, i.e. virgin or uncut forest, has special heritage value. The reason for this is partly aesthetic, partly scarcity value and partly ecological.

Old-growth forest is not necessary for all species of wildlife. Selection cut and clearfelled forest is often suitable and even superior habitat for some species (Davey 1990). Ecological problems associated with a lack of old-growth forest are therefore minimal in the jarrah forests, which are managed under a selection cutting system. It is an issue primarily of the karri forest. Mature and senescent trees in reserves and uncut zones provide for the requirements of a range of species which otherwise would be unable to live in the regrowth forest for many years. The most conspicuous among these are the hole nesters, many of which can feed in the regrowth forest while roosting and breeding in the older trees in road, river and stream zones or other reserves nearby.

Table 6 shows the distribution of old-growth forest in the southern forests. A very large proportion of the reserved forest is uncut old-growth forest. Of the 52 200 ha of karri forest types reserved, some 81.6 per cent is old-growth, 63.2 per cent of the 153 800 of jarrah types reserved is old-growth, and 68.3 per cent of the total of 211 100 ha of forest reserved is old-growth. These figures do not include the 76 100 ha of road, river and stream zoning systems, much of which is old-growth forest.

Scenic and wilderness values

The visual quality of a landscape is a value in its own right. This resource can be assessed and managed in much the same way as other resource values such as fauna, flora, water, timber and recreation (CALM 1987b).

Managing the visual resource is dependent on a knowledge and assessment of the landscape itself. Since 1987, CALM has adopted a systematic approach to inventorying and assessing landscape values based on systems now operating in other Australian States and overseas (Leonard and Hammond 1984).

The process of describing scenic values begins with a classification of broad landscape character types. A preliminary draft of landscape character types has been prepared for Western Australia (Stuart-Street and Kirkpatrick in prep).

Six landscape character types are recognized in the southern forests (see Fig. 6). All are represented in the reserve system. The landscape character type which is least represented is the Bannister uplands. This type can be found in the proposed Perup Nature Reserve and proposed nature reserves at Bokerup, Bolbelup and Unicup. This landscape character type is also well represented in reserves in forests to the north of the area (Revell[19] personal communication).

Many of the major features, scenic views, coastal landscapes and wilderness areas are incorporated into the reserve system. The special scenic areas mentioned in Chapter 2 are almost all within national parks and reserves.

Wilderness is a value which has been difficult to define. There have been numerous attempts to develop a formal definition of wilderness, but general acceptance of a single definition has not been achieved. Nevertheless, awareness of the value and scarcity of remote and natural land in Australia has grown rapidly since the 1970s (Lesslie *et al.* 1988).

Wilderness areas are generally defined as those lands in the most natural condition, with

[19] G. Revell, Department of Conservation and Land Management, W.A.

Table 6

REGENERATION STATUS OF THE FORESTED AREAS (ha) IN THE SOUTHERN FORESTS (as at 31 December 1989)

FOREST TYPE	[a] NATIONAL PARK				[a] NATURE RESERVE				[a] CONSERVATION LAND				[a] MULTIPLE USE FOREST			
	SC[b]	CF[c]	Old-Growth	Total	SC	CF	Old-Growth	Total	SC	CF	Old-Growth	Total	SC	CF	Old-Growth	Total
JARRAH-TINGLE	-	-	1 800	1 800	-	-	-	-	-	-	-	-	-	100	1 900	2 000
JARRAH	12 000	500	61 500	74 000	41 900	-	24 600	66 500	9 900		1 600	11 500	418 100	18 400	205 200	641 700
PURE KARRI	1 700	2 300	12 500	16 500	-	-	400	400	200	100	200	500	12 300	16 400	11 700	40 400
KARRI-TINGLE	-	400	2 900	3 300	-	-	-	-	-	-	-	-	-	600	1 600	2 200
KARRI	4 800	2 200	22 600	29 600	300	-	900	1 200	300		400	700	23 800	22 900	33 500	79 200
WANDOO	-	-	-	-	400	-	200	600	-	-	-	-	100	-	-	100
OTHER SPECIES	-	-	4 200	4 200	100		100	200	100			100	27 800	300	3 100	31 200
TOTAL	18 500	5 400	105 500	129 400	42 700	-	26 200	68 900	10 500	100	2 200	12 800	482 100	58 700	256 000	796 800

[a] Total areas as proposed by CALM (1987b)
[b] SC - Selection cut and regenerated
[c] CF - Clearfelled - regeneration completed or commenced

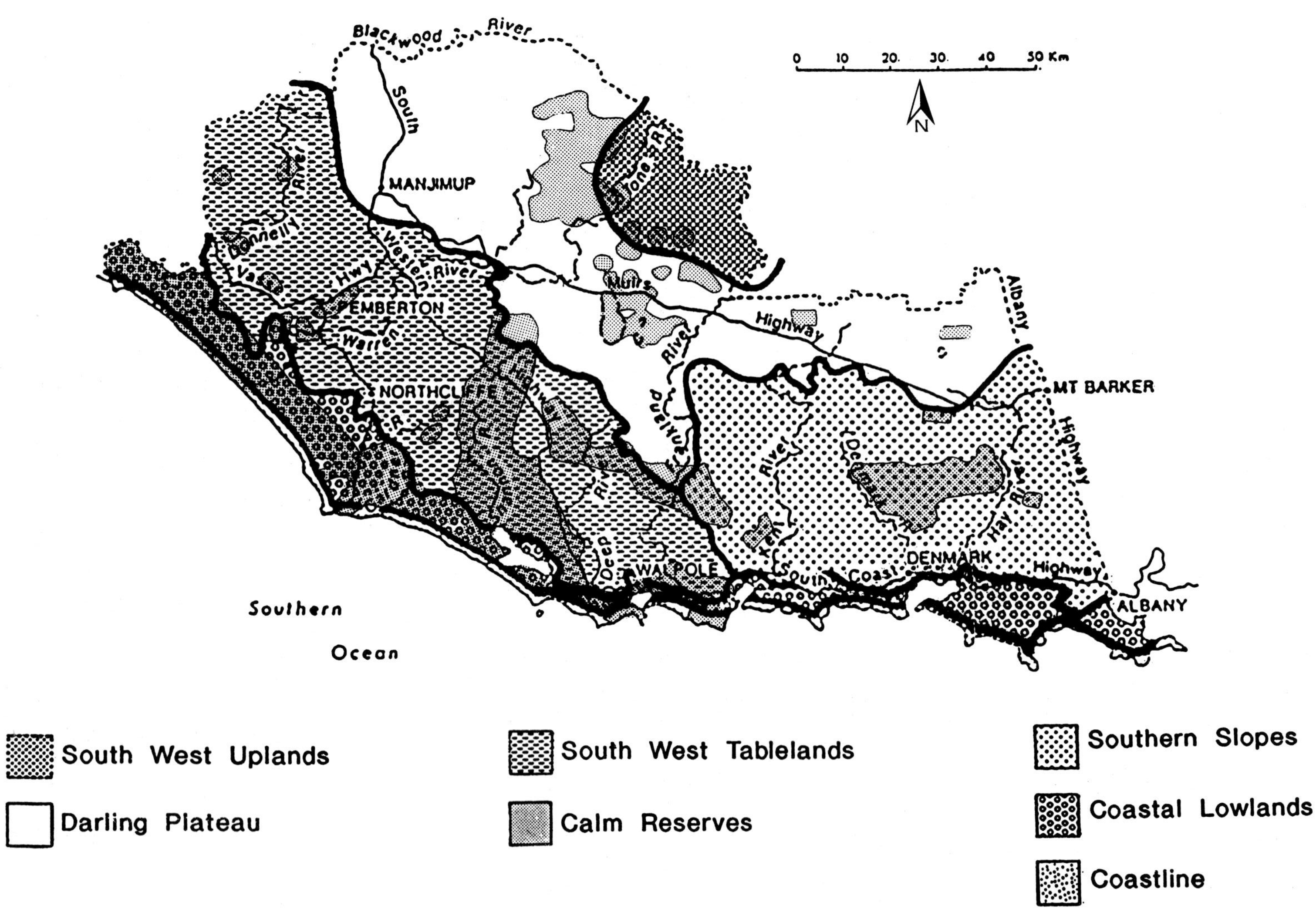

Figure 6
Broad landscape character types in a portion of the southern forests (after Stuart-Street and Kirkpatrick in prep).

minimal impacts of technological society. Wilderness areas are characterized by their remoteness and naturalness, but may reflect in varying degrees changes to the environment brought about by 40 000 or more years of Aboriginal use (Lesslie *et al.* 1988).

A method of wilderness evaluation was used by Lesslie *et al.* (1988) to develop an inventory of wilderness quality in Tasmania. The methodology, based on the wilderness continuum concept (Lesslie and Taylor 1983), places emphasis on measuring variation in wilderness quality using the following indicators.

- Remoteness from settlement: remoteness from settled (cleared) land or, within natural areas, points of permanent occupation.
- Remoteness from access: remoteness from constructed vehicle access routes.
- Aesthetic naturalness: the degree to which the landscape is free from the presence of the permanent structures of modern technological society.
- Biophysical naturalness: the degree to which the natural environment is free of biophysical disturbance owing to the influence of modern technological society.

By combining wilderness indicator values, maps can be generated showing the distribution of wilderness quality. A preliminary assessment of wilderness values in CALM's Southern Forest Region has been conducted using the indicator values developed by Lesslie (Walker[20] personal communication) (Fig. 7).

This assessment revealed that the majority of areas evaluated as having moderate, high, or very high wilderness quality located west of the Frankland River are in existing or proposed reserves. East of the Frankland River areas assessed to have moderate to high wilderness quality are located outside reserves in State forest or in vacant Crown land proposed to become State forest (see Fig. 7). The virgin status of these forests and the existence of fewer roads and tracks contributed to the high wilderness quality rating. This area, between the Frankland and the Hay Rivers, also contains extensive tracts of low open forest and low woodland of jarrah, one of the plant associations listed earlier as being poorly represented in the reserve system, and recommended for reservation.

Elsewhere in the area, the intensive development of roads and tracks, the extent of past grazing and logging, and the proximity of cleared private land, have contributed to the rating of wilderness values as low or below threshold.

Historic Values

Aboriginal

In her book *Fire and Hearth*, Hallam (1975) provides circumstantial evidence that Aborigines seldom frequented the main forest belt. They used the resource rich areas to the east and the coastal strip where access was better and resources were easier to obtain.

The distribution of Aboriginal artefacts and other evidence of their presence gives some support to this hypothesis (see Chapter 3). Most signs of their past presence are along the coastal fringe. The area to the east is now cleared farmland. There are few recorded Aboriginal sites within the main forest belt. A very large portion of the coastal area is reserved, but further effort is still needed to ensure that valuable sites are protected.

Much of the coastal strip is sand, which over the last few thousand years has moved many times, so valuable material undoubtedly still lies beneath the surface in places. Good examples of this are the recent discoveries beneath the waters of Lake Jasper.

[20] A. Walker, Department of Conservation and Land Management, W.A.

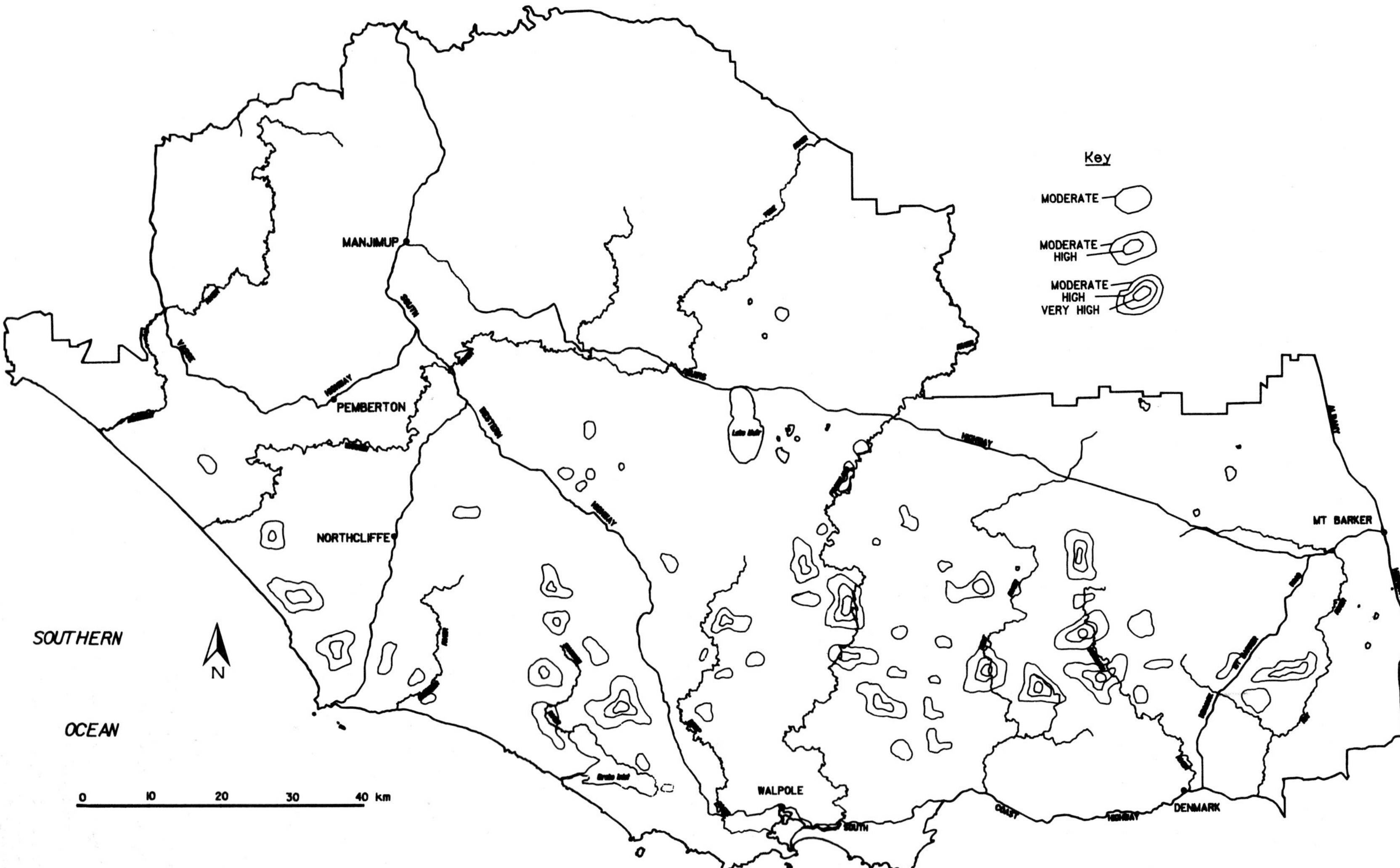

Figure 7
Wilderness quality map of CALM's Southern Forest Region (CALM 1987b) developed by A. Walker, CALM.

European

In the short history of European settlement of the south-west, a surprising amount of activity has taken place in the area. Mining, early timber cutters and sawmills, farmers and fishers have all made their mark and some relics remain today. CALM is currently making an inventory, along with other bodies, enumerating items and sites of historic value.

Fire

Fire deserves a special mention. It has been a major factor in shaping the present ecosystems which we are seeking to conserve. Without fire it will not be possible to ensure the continuance of the processes which for so long have shaped communities of plants and animals in the area. Prescribed fire, for it cannot be possible to have it otherwise, has a vital role to play in conservation areas. Continuing research, together with innovative management, is needed to keep improving prescriptions to ensure that fire fulfils its natural function as closely as possible within the framework of a rural community.

Distribution and Size of Reserves

The importance of the geographical distribution factor has been commented on earlier in this Chapter. One of the fundamental attributes of the Forests Department's original MPA system was adequate geographical dispersion of areas reserved primarily for conservation purposes. To this has been added the concept of some larger reserves. In fact, the current reserve system is a combination of two different approaches, a situation which arguably comes close to the ideal.

With respect to size, there are few factual scientific statistics on the population size necessary for adequate conservation of genetic variability in different species. While there is little agreement on the theory, the fact is that in practice, populations of several species, for example in North America the American bison and northern and southern elephant seals, have recovered successfully from very small numbers. Likewise, the area required for the successful long term conservation of such viable populations (if the size were known) is in some dispute, as pointed out by Havel (1989).

This is because the space requirements of a species, the home range or territory, is difficult to calculate with precision. Substantial differences occur depending on the method used to obtain the information, the quality of the habitat, condition of the habitat and other factors (see Christensen 1980b; Havel 1989). For these reasons, the fact that almost all of the forested reserves in the area are surrounded by State forest is of paramount importance. Unlike most reserves in the wheatbelt area to the north-east of the southern forests (Kitchener *et al.* 1980; Kitchener 1982) which are like islands in a sea of agricultural lands unsuitable for the majority of native animals, forest reserves are surrounded by little-altered, sympathetically managed State forest, which in turn has high indigenous nature conservation values.

Moist habitats

The southern forests are noteworthy for their flora and fauna with Gondwanan affinities (see Chapter 4). It is important that the reserve system contains the full range of sites with Gondwanan characteristics. The following description is provided by A.R. and B.Y. Main (personal communicaiton).

Gondwana characteristics. The chain of soil types (the catenas) of the valley sides, even in saline areas, often have a series of summer dry swamps each with its own characteristic fauna. Better drained areas permit the growth of karri and jarrah which provides a forest environment and another Gondwana-like habitat. Thus each land surface has its own hydrological regime and the potential to retain different Gondwanic elements. The retention of whole catenas with perched swamps and impeded groundwater flow extending from rocky granitic crests to valley floors is important. However, the poor drainage and shallow gradients offer great potential for the spread of phytophthora. This risk implies that, where possible, selected areas should be duplicated with at least some within established quarantine areas. Because of the shallow gradients on the incipient lateritic etch plains selected areas should be large.

Habitat Indicators

A combination of mild equable climate, relatively high moisture availability and the presence of a closed canopy provided by a suite of species representing at least some of the following families, Myrtaceae, Protaceae, Casuarinaceae, Epacridaceae, Goodeniaceae, Rutaceae, Stylideaceae and Restionaceae provides the habitat requirements of most Gondwanic invertebrate species, and their retention is essential for successful conservation. They are at risk from canopy destruction, e.g. clearing, too frequent burning, disease (*Phytophthora*), or drainage.

Summary

Gondwanic elements would be expected to persist and be found if searched for in:

1. High rainfall areas with hot summer drought;
2. Topographically high coastal areas subject to onshore south-east winds and drizzle;
3. Areas adjacent to granitic rocks from which water is shed;
4. Areas of impeded ground water flow producing winter-wet swamps;
5. Streams with extensive head waters and year round flow;
6. Areas where vegetation can harvest water from fog or cloud by drip from leaves and stem flow, e.g. tingle forest and coastal dunes and heaths;
7. Areas with southern or south-western aspect and thus sheltered from summer insolation; and
8. Because of risks from salination, aquatic and terrestrial animals need to be considered separately.

The importance of small streams, swamps, permanent or seasonal and vegetation associations with dense, closed canopy, especially on the south coast, is obvious. Such habitats are well represented in the reserve system, but no qualitative assessment has been made. Likewise, there has been no assessment of the representation of whole catenas extending from granitic monadnocks to the valley floor, although most of the southern parks and reserves are known to provide a high degree of retention of such catenas.

Conservation Values of the Multiple Use Forest

The role of the forest from which timber is being harvested is critical. In the first place, it was the value of the timber which enabled the forest to escape the early onslaught of agriculture. The value of the trees in the forest continues to contribute to the conservation of other forest values.

Secondly, multiple use forests act as buffers to conservation areas. Good forest management can ensure that no species are lost in the long term from wood production. The forests continue to provide suitable habitats for the full range of plants and animals, albeit in successional series throughout their extent. Indeed, selection cutting or clearfelling in well dispersed coupes can provide diversity of habitats, adding a further dimension to conservation values in the area as a whole.

Too much attention can be paid to conservation within boundaries drawn arbitrarily on a map, with scant regard for biological values such as home ranges of the animals living there. Animals and plants do not recognize boundaries made by people, and may continue living in forest designated for multiple use, unaware of the controversy raging about them.

The management of the whole forest system, conservation reserves and State forests is important; the two components should never be regarded as separate, as each affects and complements the other.

Summary

The southern forests have a rich flora and fauna with many relictual species, especially among the lower phyla. Many ancient forms have persisted in this small wet corner of Australia. It is fortunate that few species appear to have been lost in recent times. The timber value of the forests and the low nutritional status of the non-forested land, together have ensured that only a low proportion of the area was alienated and cleared for agriculture.

The present reserve system contains representative areas of the full range of ecological types to be found within the southern forests. All major areas of special scientific value are also largely contained within the reserve system. Further reservation of ecological types which are not well represented within the system, together with some additions to the areas of scientific interest, are desirable to make the reserve system more complete. The areas concerned were listed earlier in this Chapter.

A large proportion of the rare and restricted plant and animal taxa of the area occurs within reserves. Small additions to the reserve system may be desirable to include further species if new information on their conservation status comes to hand.

Old-growth forest is well represented within the reserve system. Likewise, the majority of features of exceptional scenic value are believed to be within the reserve system. Extensive vistas need protection, for example in parks like Mt. Frankland, where careful landscape planning to maintain magnificent vistas across forests and swamps must be carried out. Wilderness values are well catered for, but there is scope for increases through the addition of areas of low open forest and woodland of jarrah on the Kent River and further additions to the Sheepwash-Mt. Lindesay National Park. These areas have also been recommended here for their scientific values. Most of the known archaeological and Aboriginal sites of significance are within the reserve system. Some possible sites on the south coast known to the Western Australian Museum, which have yet to be investigated, may be outside reserves.

Finally, it is considered that there is good regional distribution of reserves, which together with the reserved zones along rivers, streams and roads and the woodland, heathland, sedgeland and rock outcrop reserves, forms a well integrated system. The system needs to be carefully assessed for full representation of the range of moist habitats (Gondwanan relict sites). It is important that a range of whole catenas from rocky granitic crests to the valley floor are included. There may also be small areas, moist sites, in the karri forest which could be important to preserve undisturbed.

The enormous advantage of having the entire reserve system buffered by multiple use forests managed sympathetically for nature conservation, cannot be overstated. For example, the strict dieback disease hygiene measures which are being enforced throughout the southern forests are of critical importance to the long term survival of many species of flora and fauna. The value of the multiple use forest for conservation should also not be lost sight of. It has, after all, protected the ecological values of the area very well to date.

Options in the future

This review has highlighted several plant formations which are poorly represented in the reserve system. The opportunity should be taken to examine these more closely, and to set aside representative parts for inclusion in the reserve system. There is also scope for addition to the wilderness values of the reserve system as suggested earlier.

Finally, there is flexibility for smaller changes to the system. Use of the multiple use forests for timber harvesting provides access to parts of the forest which were previously inaccessible. This process sometimes uncovers small areas significant for conservation, or scenic and other related values which were not previously recognized. In tall forest such areas are not easily identified in advance because of the difficulty of access. Continuing research will uncover new species and improve our knowledge of ecosystems; this new information may require changes to the reserve system. A system of flexible management which allows for reallocation of a portion of the buffer zones to such areas of special significance is desirable. There are also options to lengthen rotations in second growth forests, thereby improving their value for all uses.

With the additions which have been suggested, it is considered that the present reserve system is sufficient and well balanced with the other land uses in the area such as timber growing and agriculture. The total reserve system, together with the areas used primarily for timber production, comprise a conservation unit that will provide security for scientific values as well as catering for social, cultural and spiritual values into the future.

REFERENCES

ABBOTT, I. (1982). The vascular flora of the Porongurup Range of south-western Australia. *W.A. Herbarium Research Notes* **7**, 1-16.

ABBOTT, I. (1990). Insect outbreaks in forests of Western Australia. In *Population dynamics of forest insects*. Eds. A.D. Watt, S.R. Leather, M.D. Hunter and N.A. Kidd. Intercept, Andover, Hampshire. pp 95-103.

ABBOTT, I. (1992). Ecological implication of insect pests in jarrah and karri forest. In: Research on the Impact of Forest Management in south-west Australia. Department of Conservation and Land Management (W.A.) Occasional Paper 2/92.

ABBOTT, I. (In press). Prodromus of the occurrence and distribution of insect species in the forested part of south-west Western Australia. Department of Conservation and Land Management (W.A.) Technical Report.

ABBOTT, I., SMITH, R., WILLIAMS, M. and VOITIER, R. (1991). Infestation of a regenerated stand of karri (*Eucalyptus diversicolor*) by bullseye borer (*Trycopharia acanthocera*), Cerambycidae in Western Australia. *Australian Forestry* **54**, 66-74.

ALLEN, G.R. (1982). *A Field Guide to Inland Fishes of Western Australia*. W.A. Museum, Perth.

ANON. (1969). *50 Years of Forestry in Western Australia*. Forests Department (W.A.) Perth.

ANON. (1975). *A Natural System of Ecological Reserves in Australia*. Australian Academy of Science (Canberra), Report No. 19.

ANON. (1980). Report by the Steering Committee for Research and Land Use and Water Supply on Research into the Effects of the Woodchip Industry on Water Resources in South-Western Australia. Department of Conservation and Environment (W.A.) Bulletin No. 81.

ANON. (1986). Tropical Rainforests of North Queensland: Their Conservation Significance. A report by the Rainforest Conservation Society of Queensland to the Australian Heritage Commission. Australian Heritage Commission Special Australian Heritage Publication Series No. 3, Australian Government Publishing Service, Canberra.

ANON. (1987a). The impact of logging on the water resources of the southern forests, W.A. A report by the Steering Committee for Research on Land Use and Water Supply. Water Authority of W.A. Report No. WH41.

ANON. (1987b). Forest management to increase water yield from the northern jarrah forest. A report by the Steering Committee for Research on Land Use and Water Supply. Water Authority of W.A. Report No. WS3.

ANON. (1987c). *Timber Production in Western Australia: A strategy to take W.A'.s south-west forests into the 21st Century*. Department of Conservation and Land Management, Como, W.A.

ANON. (1989). Stream salinity and its reclamation in south-west Western Australia. Report of the Steering Committee for Research on Land Use and Water Supply. Water Authority of W.A. Report No. WS52.

ANON. (1990). Nature conservation, landscape and recreation values of the Lesueur area. Report to EPA from Department of Conservation and Land Management. Environmental Protection Authority Bulletin 424.

ASHTON, D.H. (1981). Fire in tall open forests (wet sclerophyll forests). In *Fire and the Australian Biota*. Eds. A.M. Gill, R.H. Groves and I.R. Noble. Australian Academy of Science, Canberra. pp 339-366.

ATTIWILL, P.M. (1982). Karri forest conservation: report and recommendations by the EPA. A report on a consultancy to the Environmental Protection Authority (W.A.). Department of Conservation and Environment (W.A.) Bulletin No. 123, Appendix A.

AUSTIN, C.M. (1986). Electrophoretic and morphological systematic studies of the genus *Cherax* (Decapoda, Parastacidae) in Australia. PhD Thesis, Zoology Department, University of W.A., Nedlands.

AUSTIN, M.P. and MILLER, D.J. (1978). Conservation. In: *Land use on the south coast of New South Wales.* Eds P.M. Fleming and J. Stokes. Vol. 4: Land function studies. CSIRO, Melbourne pp. 169-195 .

AXELROD, D.J. and RAVEN, P.H. (1982). Paleobiogeography and origin of the New Guinea flora. In *Ecology and Biogeography in New Guinea.* Ed. L. Gressett, Vol.1, Junk, The Hague pp. 919-943.

BALME, J. (1979). Artificial bias in a sample of kangaroo incisors from Devil's Lair, W.A. *Records of the Western Australian Museum* 7 (2), 229-224.

BALME, J., MERRILEES, D., and PORTER, J.K. (1978). Late Quaternary mammal remains, spanning about 30 000 years from excavation in Devil's Lair, W.A. *Journal of the Royal Society of Western Australia* **61,** 33-65.

BARENDSE, W. (1984). Speciation in the genus *Crinia* (Anura: Myobatrachidae) in southern Australia: a phylogenetic analysis of allozyme data supporting endemic speciation in south-western Australia. *Evolution* **38**, 1238-1250.

BEADLE, N.C.W. (1954). Soil phosphate and the delineation of plant communities in eastern Australia. *Ecology* **35,** 370-375.

BEADLE, N.C.W. (1966). Soil phosphate and its role in molding segments of the Australian flora and vegetation with special reference to xeromorphy and sclerophylly. *Ecology* **47**, 991-1007.

BEADLE, N.C.W. (1981). Origins of the Australian angiosperm flora. In *Ecological Biogeography of Australia.* Ed A. Keast. Junk, The Hague pp. 407-426.

BEARD, J.S. (1980). A new phytogeographic map of Western Australia. *W.A. Herbarium Research Notes* **3,** 37-58.

BEARD, J.S. (1981). The vegetation of the Swan Area. Explanatory notes to Sheet 7 of *Vegetation Survey of Western Australia: Swan.* University of W.A. Press, Nedlands.

BENZIE, J.A.H. (1987). The biogeography of Australian *Daphnia* (Cladocera, Daphniidae) : clues of an ancient origin (greater than 70 m.y. for the genus). *Hydrobiologia* **145,** 51-65.

BISHOP, M.J. (1981). The biogeography and evolution of Australian land snails. In: *Ecological Biogeography of Australia.* Ed. A. Keast Junk, The Hague pp. 923-954.

BLACKALL, W.E. AND GRIEVE, B.J. (1954-1988). *How to know Western Australian Wildflowers.* Parts I-IV. University of W.A. Press, Nedlands.

BLAKERS, M., DAVIES, S.J.J.F. and REILLY, P.N. (1984). *The Atlas of Australian Birds.* RAOU. Melbourne, University Press, Vic.

BORG, H., KING, P.D. and LOH, I.C. (1987a). Stream and ground water response to logging and subsequent regeneration in the southern forest of Western Australia: interim results from paired catchment studies. Water Authority of W.A. Report No. WH34.

BORG, H., STONEMAN, G.L. and WARD, C.G. (1987b). Stream and ground water response to logging and subsequent regeneration in the southern forest of Western Australia: results from four catchments. Department of Conservation and Land Management (W.A.) Technical Report 16.

BOULTON, A.J. and KNOTT, B (1984). Morphological and electrophoretic studies of the palaemonidae (Crustacea) of the Perth region, Western Australia. *Australian Journal of Marine Freshwater Research* **35,** 769-783.

BOUSFIELD, L.R. and JAMES, S.H. (1976). The behaviour and possible cyto-evolutionary significance of B chromosomes in *Dampiera linearis* (Angiospermae:Goodeniaceae). *Chromosoma* (Berl.) **55**, 309-323.

BRADSHAW, J. (1992 in press). Quantifying edge effects and patch size for multiple use silviculture - a discussion paper. *Forest Ecology and Management.*

BRADSHAW, F.J. and LUSH, A.R. (1981). *Conservation of the Karri forest.* Forests Department of W.A. Perth.

BRADSHAW, F.J., UNDERWOOD, R.J., QUAIN, S.J., WHITE, B.T. and GRACE, D.E. (1975). *Proposal for a South Coast National Park.* W.A. Division of The Institute of Foresters of Australia, Perth.

BREIDAHL, R. and HEWETT, P. (1992). A review of silviculture research in the karri *(Eucalyptus diversicolor)* forest. Department of Conservation and Land Management (W.A.) Research Bulletin 5.

BRIDGEWATER, P.B. (1981). Potential application of the Zurich-Montpellier system of vegetation description and classification in Australia. In *Vegetation classification in Australia.* Eds A.N. Gillison and D.J. Anderson. CSIRO, Canberra. pp. 1-9.

BURBIDGE, A.A. (1967). The biology of the south-west Australian tortoises. Ph.D Thesis University of W.A., Nedlands.

BURBIDGE, A.A. (1981). The ecology of the western swamp tortoise *Pseudemydura umbrina* (Testudines): (Chelidae). *Australian Wildlife Research* **8**, 203-223.

BURBIDGE, A.A. and MacKENZIE, N.L. (1989). Pattern in the modern decline of Western Australian vertebrate fauna: causes and conservation implications. *Biological Conservation* **50**, 143-198.

BURBIDGE, A.A, FOLLEY, G.L. and SMITH, G.T. (1986). The Noisy Scrub-Bird. Department of Conservation and Land Management (W.A.) Wildlife Management Program No. 2.

BURBIDGE, A.A., KIRSCH, J.A. and MAIN, A.R. (1974). Relationships within the Chelidae (Testudines: Pleurodira) of Australia and New Guinea. *Copeia* **2**, 392-409.

BURBIDGE, N.T. (1960). The phytogeography of the Australian region. *Australian Journal of Botany* **8**, 75-212.

BUREAU OF METEOROLOGY (1965). Climate Survey Region 16 - South-West Western Australia. Issued by the Director or Meteorology, Melbourne.

BUREAU OF METEOROLOGY (1975). *Climate averages Western Australia.* Australian Government Publishing Service, Canberra.

BURGMAN, M.A. (1988). Spatial analysis of vegetation patterns in southern Western Australia: implications for reserve design. *Australian Journal of Ecology* **13**, 415-429.

CALM (1987a). *Regional Management Plan 1987 - 1997, Central Forest Region.* Department of Conservation and Land Management, Como, W.A.

CALM (1987b). *Regional Management Plan 1987 - 1997, Southern Forest Region.* Department of Conservation and Land Management, Como, W.A.

CALM (1987c). *Regional Management Plan 1987 - 1997, Northern Forest Region.* Department of Conservation and Land Management, Como, W.A.

CALM (1987d). *Shannon Park and D'Entrecasteaux National Park Management Plan.* Department of Conservatioin and Land Management, Como, W.A.

CALM (1988). Annual Report 1987/88 - Department of Conservation and Land Management, Como, W.A.

CALM (1990a). *Walpole-Nornalup National Park Management Plan.* Department of Conservation and Land Management, Como, W.A.

CALM (1990b). *Manual of logging specifications for management of hardwood and softwood logging operations in the south-west of Western Australia.* 3rd edn. Department of Conservation and Land Management, Como, W.A.

CARLQUIST, S. (1974). Adoptive radiation: the flora of southwestern Australia. In *Island Biology*. Chapter 8. Columbia University Press, New York. pp. 253-331.

CHRISTENSEN, P. (1973). A new concept in forestry - fauna priority areas. Forests Department (W.A.) *Forest Focus* No.10.

CHRISTENSEN, P. (1974). The concept of fauna priority areas. In *Third Fire Ecology Symposium*. Monash University and Forests Commission Victoria pp. 66-73.

CHRISTENSEN, P. (1980a). *A Forest Worth Preserving*. Unpublished, Forests Department of W.A, Perth.

CHRISTENSEN, P.E.S. (1980b). The biology of *Bettongia penicillata* Gray, 1837, *and Macropus eugenii* (Desmarest, 1917) in relation to fire. Forests Department (W.A.) Bulletin 91.

CHRISTENSEN, P. (1982). The distribution of *Lepidogalaxias salamandroides* and other small freshwater fishes in the lower south-west of Western Australia. *Journal of the Royal Society of Western Australia* **65** (4), 131-141.

CHRISTENSEN, P. (1987a). Forest wildlife and fire in Australian eucalypt forests. In Proceedings 18th IUFRO World Conference, Ljubljana, Yugoslavia.

CHRISTENSEN, P. (1987b). The Perup Forest, Western Australia. In: *Forest Management in Australia*. Proceedings 1987 Conference, Institute of Foresters of Australia, Perth, W.A. pp. 373-385.

CHRISTENSEN, P. and ABBOTT, I. (1989). Impact of fire in the eucalypt forest ecosystems of southern Western Australia: a critical review. *Australian Forestry* **52** (2), 103-121.

CHRISTENSEN, P. and ANNELS, A.R. (1985). Fire in southern tall forests. In *Fire Ecology and Management of Western Australian Ecosystems*. Proceedings of a Symposium, May 1985. Ed. J.R. Ford. WAIT Environmental Studies Group Report No.14, Western Australian Institute of Technology, Perth. pp 67-82.

CHRISTENSEN, P. and BURROWS, N. (1985). Fire, an old tool with a new use. In *Ecology of Biological Invasions.: an Australian Perspective*. Eds. R.H. Groves and J.J. Burton. Australian Academy of Science, Canberra. pp. 35-42

CHRISTENSEN, P.E. and KIMBER, P.C. (1975). Effect of prescribed burning on the flora and fauna of south-west Australian forests. *Proceedings of the Ecological Society of Australia* **9**, 85-106.

CHRISTENSEN, P. and MAISEY, K. (1987). The use of fire as a management tool in fauna conservation reserves. In: *Nature Conservation: The Role of Remnants of Native Vegetation*. Eds D.A. Saunders, G.W. Arnold, A.A. Burbidge and A.J.M. Hopkins. Surrey Beatty and Sons, Chipping Norton, N.S.W. pp. 15-28.

CHRISTENSEN, P., ANNELS, A., LIDDELOW, G., and SKINNER, P. (1985). Vertebrate fauna in the southern forests of Western Australia: a survey. Forests Department (W.A.) Bulletin 94.

CHRISTENSEN, P., MAISEY, K. and PERRY, D.H. (1984). Radiotracking the numbat, *(Myrmecobius fasciatus)*, in the Perup Forest of Western Australia. *Australian Wildlife Research* **11**, 275-288.

CHRISTENSEN, P., PENTONY, K. and SCHMIDT, W. (1981). The Blackwood: a valley in transition. Forests Department (W.A.) *Special Focus No 2*.

CHRISTENSEN, P.E.S., WARDELL-JOHNSON, G.W. and KIMBER, P. (1986). Birds and fire in southwestern forests. In *Birds of Eucalypt Forest and Woodlands: Ecology, Conservation, Management*. Eds A. Keast, H.F. Recher, H. Ford and D. Saunders. Surrey, Beatty and Sons, Chipping Norton, N.S.W. pp. 291-299.

CHURCHILL, D.M. (1968). The distribution and prehistory of *Eucalyptus diversicolor* F. Muell. in relation to rainfall. *Australian Journal of Botany* **16**, 125-152.

CHURCHWARD, H.M, McARTHUR, W.M, SEWELL, P.L, and BARTLE, G.A. (1988). Landforms and soils of the southcoast and hinterland, Western Australia. CSIRO Division of Water Resources Divisional Report 88/1.

CLARKE, J. (1983). An Aboriginal engraving site in the south-west of Western Australia. *Records of the Western Australian Museum* **11,** 63-67.

COATES, D.J. and R.E. SOKOLOWSKI (1989). Geographic patterns of genetic diversity in karri (*Eucalyptus diversicolor* F. Muell.). *Australian Journal of Botany* **37,** 145-156.

COCKBAIN, A.E. (1968). The stratigraphy of the Plantagenet Group, Western Australia. W.A. Geological Survey Annual Report pp. 61-63.

COLLINS, P.D.K. and BARRETT, D.F. (1980). Shannon, Warren and Donnelly River basins water resources survey. W.A. Public Works Department, Water Resources Branch Report No. 6.

COLLINS, P.D.K. and FOWLIE W.G. (1981). Denmark and Kent River basins water resources survey. W.A. Public Works Department, Water Resources Branch Report.

COMMITTEE OF INQUIRY ON THE SHANNON RIVER BASIN (1987). Report to the Hon Minister for Conservation and Land Management. Department of Conservation and Land Management, Perth, W.A.

COMMON, I.F.B. (1990). *Moths of Australia.* Melbourne University Press, Vic.

CONSERVATION COUNCIL OF W.A. (1982). *Karri at the Crossroads.* Campaign to save Native Forests, South-West Forests Defence Foundation, Australian Conservation Foundation (Inc) Western Chapter.

COPE, R.N. (1975). Tertiary epeirogeny in the southern part of Western Australia. W.A. Geological Survey Annual Report 1974 pp. 40-46.

CRAWFORD, I.M. (1972). Function and change in Aboriginal rock art, Western Australia. *World Archaeology* **3,** 301-312.

CTRC (1974). Conservation Reserves in Western Australia. Report of the Conservation Through Reserves Committee to the Environmental Protection Authority. Systems 1-5 and 8-12.

CURRY, S.J., HUMPHREY, C.W.F., KOCH, L.E. and MAIN, B.Y. (1985). Changes in Arachnid communities resulting from forestry practices in karri forest, south-west Western Australia. *Australian Forestry Research* **15,** 469-480.

DAMES AND MOORE (1982). *Environmental Assessment: Proposed Alternative Dam Sites Lefroy Brook and Four Mile Brook, Manjimup, W.A.* Dames & Moore, Perth.

DAMES AND MOORE (1985). Public environmental report Pemberton trout hatchery water supply Big Brook Dam. Engineering Division W.A. Public Works Department Report No. WRB 117

DARRAGH, T.A. and KENDRICK, G.W. (1980). Eocene bivalves from the Pallingup siltstone near Walpole, Western Australia. *Journal of the Royal Society of Western Australia* **63,** 5-20.

DAVEY, S.M. (1990). The environmental relationships of arboreal marsupials in a eucalypt forest: a basis for Australian wildlife management. PhD Thesis, Australian National University, Canberra.

DAVIDSON, D.S. (1952). Notes on the pictographs and petroglyphs of Western Australia and a discussion of their affinities with appearances elsewhere on the continent. *Proceedings of the American Philosophy Society* **96,** 76-117.

DAYTON, L. (1990). New life for old forest. *New Scientist* (October 1990) pp. 21-25.

DELL, B., HAVEL, J.J. and MALAJCZUK, N. (1989). *The Jarrah Forest: A complex mediterranean ecosystem.* Kluwer Academic Publishers, Dordrecht, The Netherlands.

DIELS, L. (1906). *Die pflanzenwelt von West Australien sudlich des Wedeskreises.* Engelmann, Leipzig.

DIMMOCK, G.M, BETTENAY, G. and MULCAHY, M.J. (1974). Salt content of lateritic profiles in the Darling Range, Western Australia. *Australian Journal of Soil Research* **12,** 63-69.

DIX, W.C. and MEAGHER, S.J. (1976). Fish traps in the south-west of Western Australia. *Records of the Western Australian Museum* **4**(2), 171-187.

DORTCH, C.E. (1974). A twelve thousand year old occupation floor in Devil's Lair, Western Australia. *Mankind* **9**, 195-205.

DORTCH, C.E. (1975). Geometric microliths from a dated archeological deposit near Northcliffe W.A. *Journal of the Royal Society of Western Australia* **58**, 59-63.

DORTCH, C.E. (1976). Two engraved stone plaques of Late Pleistocene age from Devil's Lair, Western Australia. *Archaeology and Physical Anthropology in Oceania* **11**, 32-44.

DORTCH, C.E. (1979a). Devil's Lair, an example of prolonged cave use in south-western Australia. *World Archaeology* **10**, 258-279.

DORTCH, C.E. (1979b). 33,000 year old stone and bone artifacts from Devil's Lair, Western Australia. *Records of the Western Australian Museum* **7**, 329-367.

DORTCH, C.E. (1980). A possible pendant of Marl from Devils Lair, Western Australia. *Records of the Western Australian Museum* **8**(3), 401-403.

DORTCH, C.E. and GARDNER, G. (1976). Archaeological investigations in the Northcliffe district, Western Australia. *Records of the Western Australian Museum* **4**, 257-294.

DORTCH, C.E, KENDRICK, W. AND MORSE, K. (1984). Aboriginal Mollusc exploitations in southwestern Australia. *Archeology in Oceania* **19**(3), 81-104.

DOW, D.B. (1977). A geological synthesis of Papua New Guinea. Bureau of Mineral Resources,Geology and Geophysics, Bulletin No. 201.

EPA (1976). *Conservation Reserves for Western Australia as Recommended by the Environmental Protection Authority: System 1, 2, 3 and 5.* Environmental Protection Authority, Perth.

EPA (1982). Karri forest conservation: report and recommendation by the Environmental Protection Authority, Department of Conservation and Environment (W.A.) Bulletin 123.

ERIKSON, R., GEORGE, A.S., MARCHANT, N.G., AND MORCOMBE, M.K. (1973). *Flowers and Plants of Western Australia.* Reed, Sydney.

FAEGRI, K. (1965). Reflections on the development of pollination systems in African Proteaceae. *Journal of South African Botany* **31**, 133-136.

FERGUSEN, W.C. (1981). Archaeological investigations at the Quininup Brook site complex, Western Australia. *Records of the Western Australian Museum* **8** (4), 609-637.

FORD, J.R. (ed.) (1985). *Fire Ecology and Management of Western Australian Ecosystems.* Proceedings of a Symposium WAIT Environmental Studies Group Report No.14, Western Australian Institute of Technology, Perth.

FORESTS DEPARTMENT OF WESTERN AUSTRALIA (1972). *General Working Plan No. 85.* Forests Department, Perth, W.A.

FORESTS DEPARTMENT OF WESTERN AUSTRALIA (1977). *General Working Plan No. 86.* Forests Department, Perth, W.A.

FORESTS DEPARTMENT OF WESTERN AUSTRALIA (1982). *General Working Plan No. 87.* Forests Department, Perth, W.A.

FORSTER, R.R. and PLATNICK, N.I. (1985). A preview of the austral spider family Orsolobidae (Arachnida, Araneae), with notes on the superfamily Dysderroidea. *Bulletin of the American Museum of Natural History* **181** (1), 1-229.

FREWER, P., HILDER, D., TAYLOR, N., BATCHELOR, M., and SIMPKIN-BROWN, J. (1987). *Leeuwin-Naturaliste National Park Draft Management Plan.* Department of Conservation and Land Management (W.A.) Como.

FRIEND, J.A. (1990). The numbat *Myrmecobius fasciatus* (Myrmecobiidae): history of decline and potential for recovery. *Proceedings of the Ecological Society of Australia.* **16,** 369-377.

GALLOWAY, R.W., and KEMP, E.M. (1981). Late Cainozoic environments in Australia. In *Ecological Biogeography of Australia.* Ed. A. Keast. Junk, The Hague. pp. 51-80.

GARDNER, C.A. (1942). The vegetation of Western Australia with special reference to climate and soils. *Journal and Proceedings of Royal Society of Western Australia* **28,** 11-87.

GARDNER, C.A. and BENNETTS, H.W. (1956). *The Toxic Plants of Western Australia.* W.A. Newspapers Ltd., Perth.

GILL, A.M. (1975). Fire and Australian flora: a review. *Australian Forestry* **38,** 4-25.

GILL, A.M., GROVES, R.H. and NOBLE, I.R. (1981). *Fire and the Australian Biota.* Australian Academy of Science, Canberra.

GILL, E.D. (1973). Prehistoric wildfires in S.E. Australia. *Victorian Naturalist* **90,** 347-348.

GLOVER, J.E. (1975). Aboriginal chert artifacts probably from quarries on the continental shelf, W.A. *Search* **6,**392-394.

GOOD, R. (1974). *The Geography of Flowering Plants.* 4th edn. Longman, London.

GREEN, J.W. (1985). *Census of the Vascular Plants of Western Australia.* Department of Agriculture (W.A.) Perth.

GRIFFIN, E.A. (1985). *Vegetation Survey of Smiths Brook Nature Reserve.* E.A. Griffin & Associates, Perth.

GRIFFIN, E.A and HOPKINS, A.J.M. (1985). The flora and vegetation of Mt. Lesueur, Western Australia. *Journal of the Royal Society of Western Australia* **67** (2), 45-57.

GROVE, T.S. and MALAJCZUK, N. (1985). Biomass production by trees and understorey shrubs in an age-series of *Eucalyptus diversicolor* F. Muell. stands. *Forest Ecology and Management* **11,** 59-74.

HALL, N., JOHNSTON, R.D. and CHIPPENDALE, G.M. (1975). *Forest Trees of Australia.* Australian Government Publishing Service, Canberra.

HALLAM, S.J. (1972). An archaeological survey of the Perth area, Western Australia : a progress report on art and artefacts, dates and demography. *Australian Institute of Aboriginal Studies Newsletter* **3** (5), 11-19.

HALLAM, S.J. (1975). *Fire and Hearth.* Australian Institute of Aboriginal Studies, Canberra.

HALLAM, S.J. (1977). Topographic archaeology and artifactual evidence. In *Stone tools as cultural makers.* Ed. R.V.S. Wright. Australian Institute of Aboriginal Studies, Canberra. pp. 169-177.

HALLAM, S.J. (1981). The first Western Australians. In *A New History of Western Australia.* Ed. C.T. Stannage. University of W.A. Press, Nedlands, W.A. pp. 35-71.

HAMMOND, J.E. (1933). *Winjan's people. The Story of the South-West Australian Aborigines.* Imperial Printing, Perth.

HAVEL, J.J (1975a). Site-vegetation mapping in the northern jarrah forest (Darling Range). 1 - Definition of site-vegetation Types. Forests Department (W.A.) Bulletin 86.

HAVEL, J.J. (1975b). Site-vegetation mapping in the northern jarrah forest (Darling Range). II - Location and mapping of site-vegetation types. Forests Department (W.A.) Bulletin 87.

HAVEL, J.J. (1989). Conservation in the northern jarrah forest. In *The Jarrah Forest: a complex mediterranean ecosystem.* Eds B. Dell, J.J. Havel and N. Malajczuk. Kluwer Academic Publishers. Dordrecht, The Netherlands. pp. 379-399.

HEDDLE, E.M., LONERAGAN, O.W. AND HAVEL, J.J., (1980). Vegetation complexes of the Darling system, Western Australia. In: *Atlas of natural resources Darling system Western Australia*. Department of Conservation and Environment, (W.A.) Perth. pp. 37-74.

HEINSELMAN, M.L. (1971). Preserving nature in forested wilderness areas and national parks. In: Proceedings of the forest recreation Symposium. Ed. E.V.H. Larson. State University, New York, Syracuse, New York.

HEYER, W.R., DAUGHERTY, C.H. and MAXON, L.R. (1982). Systematic resolution of the Crinia complex of (Amphibia: Anura: Myobatrachidae). *Proceeding of the Biological Society of Washington* **95,** 423-427.

HICKLEY, L.J. and DOYLE, J.A. (1977). Early Cretaceous fossil evidence for angiosperm evolution. *The Botanical Review* **43,** 3-104.

HILLIS, W.E. and BROWN, A.G. (1978). *Eucalypts for Wood Production.* CSIRO, Adelaide.

HILTON, R.N. (1982). A census of the larger fungi of Western Australia. *Journal of the Royal Society of Western Australia* **65** (1), 1-15.

HILTON, R.N. (1988). A census of the larger fungi of Western Australia Part II. *Journal of the Royal Society of Western Australia* **70** (4), 111-118.

HOOKER, J.D. (1859). Introductory essay: Botany of the Antarctic voyage of H.M. Discovery Ship Erebus. In *Flora Tasmaniae* Vol. 3. Reeve, London.

HOPPER, S.D. (1979). Biogeographical aspects of speciation in the southwest Australia flora. *Annual Review of Ecology and Systematics* **10,** 399-422.

HOPPER, S.D. and COATES, D.J. (1990). Conservation of genetic resources in Australia's flora and fauna. *Proceedings of the Ecological Society of Australia* **16,** 567-577.

HOPPER, S.D., KEIGHERY, G.J. and WARDELL-JOHNSON, G. (1992) Flora of the karri forest and other communities in the Warren Botanical subdistrict of Western Australia. In: Research on the impact of forest management in south-west Australia. Department of Conservation and Land Management (W.A.) Occasional Paper 2/92.

HOSKING, J.S. and BURVILL, G.H. (1938). A soil survey of part of the Denmark estate, Western Australia. CSIRO Bulletin No 115.

HOW, R.A., DELL, J. AND HUMPHREYS, W.F. (1987). The ground vertebrate fauna of coastal areas between Busselton and Albany, Western Australia. *Records of the Western Australian Musuem* **13** (4), 553-574.

HOWDEN, H.F. (1981). Zoogography of some Australian Coleoptera as exemplified by the Scarabaeiodea. In *Ecological Biogeography of Australia*. Ed. A. Keast. Junk, The Hague. pp. 1007-1036.

INIONS,G., WARDELL-JOHNSON, G. and ANNELS, A. (1990a). Classification and evaluation of sites in karri (*Eucalyptus diversicolor*) regeneration. 1. Edaphic and Climatic Attributes. *Forest Ecology and Management* **32,** 117-134.

INIONS, G., WARDELL-JOHNSON, G. and ANNELS, A. (1990b). Classification and evaluation of sites in karri (*Eucalyptus diversicolor*) regeneration. 2. Floristic attributes. *Forest Ecology and Management* **32,** 135-154.

JACKSON, A. (1931). The Oligochaeta of south-western Australia. *Journal of the Royal Society of Western Australia* **17,** 71-136.

JACOBS, M.R. (1955). *Growth Habits of Eucalypts.* Forestry and Timber Bureau, Canberra.

JAMES, S.H. and HOPPER, S.D. (1981). Speciation in the Australian flora. In *The Biology of Australian Plants* . Eds J.S. Pate and A.J. McComb. University of W.A. Press, Nedlands. pp 362-391.

JOHNSON, L.A.S., and BRIGGS, B.G. (1975). On the Proteaceae - the evolution and classification of a southern family. *Botanical Journal of the Linnean Society* **70,** 83-182.

JOHNSON, L.A.S., and BRIGGS, B.G. (1981). Three old southern families - Myrtaceae, Proteaceae and Restionaceae. In *Ecological Biogeography of Australia.* Ed. A. Keast. Junk, The Hague. pp 427-470.

JOHNSTON, C.D, McARTHUR, W.M, and PECK, A.J. (1980). Distribution of soluble salts in soils of the Manjimup Woodchip Licence Area, Western Australia. CSIRO Division of Land Resource Management, Technical Paper. No 5.

JUTSON, J.T. (1914). An outline of the physical geology (Phsiography) of Western Australia. W.A. Geological Survey Bulletin 61.

KABAY, E.D. and START, A.N. (1976). Results of the search for the Potoroo in south west and south coast of Western Australia. Department of Fisheries and Wildlife, (W.A.) Report.

KEAST, A. (1981a). Distributional patterns, regional biotas, and adaptations in the Australian biota: A synthesis. In *Ecological Biogeography of Australia.* Ed. A. Keast. Junk, The Hague. pp. 1891-1998.

KEAST, A. (1981b). Origins and relationships of the Australian biota. In *Ecological Biogeography of Australia.* Ed. A. Keast. Junk, The Hague. pp. 1999-2050.

KEIGHERY, G.J. (1984a). Phytogeography of the Fabaceae of Western Australia. *Kings Park Research Notes* **8,** 24-36.

KEIGHERY, G.J. (1984b). Phytogeography of the monocotyledons of Western Australia. *Kings Park Research Notes* **8,** 39-67.

KEIGHERY, G.J. (in prep (a)). Plants of the Torndirrup National Park.

KEIGHERY, G.J. (in prep (b)). A checklist of the vascular plants of the Porongurup National Park.

KEMP, E.M. (1981). Tertiary palaeogeography and the evolution of Australian climate. In: *Ecological Biogeography of Australia.* Ed. A. Keast. Junk, The Hague. pp. 31-50.

KENNETT, J.P. (1978). The development of planktonic biogeography in the Southern Ocean during the Cenozoic. *Marine Micropalentology* **3,** 301-345.

KINNEAR, J.E., COCKSON, A., CHRISTENSEN, P. and MAIN, A.R. (1979). The nutritional biology of the ruminants and ruminant-like mammals - a new approach. *Compendium Biochemistry Physiolgy* **64A,** 357-365.

KITCHENER, D.J. (1982). Predictors of vertebrate species richness in nature reserves in the Western Australian wheatbelt. *Australian Wildlife Research* **9,** 1-8.

KITCHENER, D.J., CAPUTI, N. and JAMES, B. (1986). Revision of Australo-Papuan Pipistrellus and of Falistrellus (Microchiroptera:Vespertilionidae). *Records of the Western Australian Museum* **12,** 435-495.

KITCHENER,D.J., CHAPMAN, A. and BARRON,G. (1978). Mammals of the Northern Swan Coastal Plain. In *Faunal Studies of the Northern Swan Coastal Plain : A Consideration of Past and Future Changes.* Western Australian Museum Report for the Department of Conservation and Environment, W.A.

KITCHENER, D.J. CHAPMAN, A., DELL, J. and MUIR, B.G. (1980). Lizard assemblage and reserve size and structure in the Western Australian wheatbelt - some implications for conservation. *Biological Conservation* **17,** 25-62.

KITCHENER, D.J., STODDART, J. and HENRY, J. (1984). A toxonomic revision of the *Sminthopsis murina* complex (Marsupialia; Dasyuridae) in Australia including descriptions of four new species. *Records of the Western Australian Museum* **11** (3), 201-247.

KITCHING, R.L. (1981). The geography of the Australian Papilionoidea. In *Ecological Biogeography of Australia.* Ed. A. Keast. Junk, The Hague. pp. 977-1006.

KOMAREK, E.V. (1967). The nature of lighting fires. *Proceedings, Tall Timbers Fire Ecology Conference* **7,** 5 - 41.

KOMAREK, E.V. (1968). Lightning and lightning fires as ecological forces. *Proceedings, Tall Timbers Fire Ecology Conference* **8,** 167-197.

KOMAREK, E.V. (1972). Ancient fires. *Proceedings, Tall Timbers Fire Ecology Conference* **12,** 219-240.

KOZLOWSKI, T.T. and AHLGREN, C.E. (1974). *Fires and Ecosystems.* Academic Press, New York.

LANGE, R.T. (1982). Australian tertiary vegetation; evidence and enterpretation. In *A history of Australasian Vegetation.* Ed. J.M.B. Smith. McGraw Hill Book Company, Sydney. pp. 44-89.

LEONARD, M. AND HAMMOND, R. (1984). *Landscape Character Types of Victoria.* Forests Commission, Vic.

LESSLIE. R.G. and TAYLOR, S.G. (1983). Wilderness in South Australia: An inventory of the State's relatively high quality wilderness areas. The Centre for Environmental Studies, University of Adelaide, Occasional Paper No. 1.

LESSLIE, R.G., MACKEY, B.G. and SHULMEISTER, J. (1988). National Wilderness Inventory. Stage II. Wilderness quality in Tasmania: A report to the Australian Heritage Commission. AHC, Canberra.

LITTLEJOHN, M.J. (1981). The amphibia of mesic southern Australia: a zoogeographic perspective. *Monographiae Biologicae* **41,** 1303-1330.

LOGAN, B.W. (1968). Western Australia. In *Quaternary Shorelines Research in Australia and New Zealand.* Ed. E.D. Gill. Australian Journal of Science Vol. 31, pp. 110.

LONERAGAN, O.W. (1979). Karri (*Eucalyptus diversicolor* F. Muell) phenological studies in relation to reforestation. Forests Department (W.A.), Bulletin 90.

MACKERRAS, I.M. (1970). Composition and distribution of the fauna. In *The insects of Australia: a text book for students and research workers.* Melbourne University Press, Vic.

MAIN, A.R. (1957). Studies in Australian Amphibia: 1. The genus *Crinia* Tschudi in south-western Australia and some species from south-eastern Australia. *Australian Journal of Zoology* **5,** 30-55.

MAIN, A.R. (1965). *Frogs of southern western Australia.* W.A. Naturalists Club, Perth.

MAIN, A.R. (1987). Evolution and radiation of the terrestrial fauna. In: *Fauna of Australia.* Eds G.R. Dyne and D.W. Walton. Vol. 1a: General articles. Australian Government Publishing Service, Canberra. pp. 136-155.

MAIN, A.R., LEE, A.K. and LITTLEJOHN, M.J. (1958). Evolution in three genera of Australian frogs. *Evolution* **14,** 224-233.

MAIN, B.Y. (1976). *Spiders.* Collins, Sydney.

MAIN, B.Y. (1981). A comparative account of the biogeography of terrestrial invertebrates in Australia: some generalisations. In *Ecological Biogeography of Australia.* Ed. A. Keast. Junk, The Hague. pp. 1055-1078.

MAIN, B.Y. (1985). Further studies on the systematics of ctenizid trapdoor spiders: A review of the Australian genera (Mygalomorphae:Ctenizidae). *Australian Journal of Zoology Supplementary Series* **108,** 1-84.

MAIN, B.Y. (1987). Ecological disturbance and conservation of spiders: Implications for biogeographic relics in southwestern Australia. In *The role of invertebrates in conservation and biological survey.* Ed. J.D. Major. Proceedings of a workshop held during the 18th scientific conference of the Australian Entomological Society. Department of Conservation and Land Management, W.A. pp. 88-97.

MARCHANT, N.G. (1973). Species diversity in the south-western flora. *Journal of the Royal Society of Western Australia* **56,** 23-30.

MARTIN, A.A. (1972). Studies in Australian amphibia III. The *Limnodynastes dorsalis* complex (Anura: Leptodactylidae). *Australian Journal of Zoology* **20,** 165-211.

MARTIN, H.A. (1981). The Tertiary flora. In *Ecological Biogeography of Australia.* Ed. A. Keast. Junk, The Hague. pp. 391-406.

MARTIN, M.W. (1987). Review of the effect of logging on groundwater in the Southern Forests of Western Australia - Project 2, Paired Catchment Study. Department of Mines (W.A.). Geological Survey Record 1986/87.

McARTHUR, W.M. and BETTENAY, E. (1960). Development and distribution of soils of the Swan Coastal Plain, W.A. CSIRO Australian Soil Publication No 16.

McARTHUR, W.M. and CLIFTON, A.L. (1975). Forestry and agriculture in relation to soils in the Pemberton area of Western Australia. CSIRO Soils and Land Use series 54.

McCUTCHEON, G.S. (1978). Broadscale forest site survey techniques used in the Donnybrook Sunkland. Forests Department (W.A.) Research Paper No 48.

McCUTCHEON, G.S. (1980). Field classification of vegetation types as an aid to soil survey. Forests Department (W.A.) Research Paper No 57.

McDOWALL, R.M. (1981). The relationships of Australian freshwater fishes. In *Ecological Biogeography of Australia.* Ed. A. Keast. Junk, The Hague. pp 1251-1274.

McDOWALL, R.M. and FRANKENBURG, R.S. (1981). The galaxiid fishes of Australia. *Records of the Australian Museum* **33**(10), 443-605.

MELVILLE, R. (1973). Relict plants in the Australian flora and their conservation. In *Nature Conservation in the Pacific.* Eds. A.B. Costin and R.H. Groves. Australian National University Press, Canberra. pp. 83-90.

MELVILLE, R. (1975). The distribution of Australian relict plants and its bearing on angiosperm evolution. *Botanical Journal of the Linnean Society* **71,** 67-88.

MERRICK, J.R. and SCHMIDT, G.E. (1984). *Australian Freshwater Fishes: Biology and Management.* School of Biological Science, Macquarie University, New South Wales.

MERRILEES, D. (1979). Prehistoric rock wallabies (Marsupialia, Macropodidae, Petrogale) in the far southwest of Western Australia. *Journal of the Royal Society of Western Australia* **61,** 73-96.

MERRILEES, D., DIX, W.C., HALLAM, S.J., DOUGLAS, W.H. and BERNOT, R.M. (1973). Aboriginal man in southwestern Australia. *Journal and Proceedings of the Royal Society of Western Australia* **56,** 44-55.

MORSE, K. (1984). First record of painted aboriginal rock art in a south-western limestone cave. *Records of Western Australian Museum* **11**(2), 197-199.

MOUNT, A.B. (1964). The interdependence of the eucalypts and forest fires in Southern Australia. *Australian Forestry* **28,** 166-172.

MUELLER, J. (1970). Palynological evidence on early differentiation in Angiosperms. *Biological Review* **45,** 417-450.

MUTCH, R.W. (1970). Wildland fires and ecosystems : a hypothesis. *Ecology* **51,** 1046-1051.

NELSON, E.C. (1975). The taxonomy and ecology of *Adenanthos Labill.* (Proteaceae) in southern Australia. Ph.D Thesis. Department of Biogeography and Geomorphology, Australian National University, Canberra.

NELSON, E.C. (1978). A taxonomic revision of *Adenanthos Labill.* (Proteaceae). *Brunonia* **1** (3).

NELSON, E.C. (1981). Phytogeography of southern Australia. In *Ecological Biogeography of Australia.* Ed. A. Keast, Junk, The Hague. pp. 733-760.

NORTHCOTE, K.H. (1971). *A Factual Key for the Recognition of Australian Soils.* Rellim Technical Publication, Adelaide.

OLIVER. A.J., KING, D.R. and MEAD, R.J. (1977). The evolution of resistance to fluroacetate intoxication in mammals. *Search* **8**, 130-132.

PATERSON, C.A. (1896). Notes about the tribes inhabiting the coastal district from Geraldton to Albany, and those of territories nearest adjoining them. *Transaction of the Royal Society of South Australia* **16**, 288-291.

PORTER, J.K. (1979). Vertebrate remains from a stratified Holocene deposit in Skull cave, Western Australia. *Journal of the Royal Society of Western Australia* **61**, 109-118.

PURDIE, R.W. and SLATYER, R.O. (1976) Vegetation succession after fire in sclerophyll woodland communities in south-eastern Australia. *Australian Journal of Ecology* **1**, 223-236.

PUSEY, B.J. and EDWARDS, D.H. (1990). Limnology of the southern acid peaty flats, South-western Australia. *Journal of the Royal Society of Western Australia* **73**, *29-46*.

RAVEN, P.H. (1973). The evolution of Mediterranean floras. In *Ecological Studies, Analysis and Synthesis.* Eds F. di Castri and H.A. Mooney. Springer-Verlag, Berlin. pp. 213-224.

RAVEN, P.H. (1979). Plate tectonic and southern hemisphere biogeography. In: *Tropical Botany*. Eds. K.J. Larsen and L.B. Holm-Nielsen. Academic Press, London. pp. 3-24.

RAVEN, P.H. and AXELROD, D.J. (1974). Angiosperm Biogeography and past continental movements. *Annals of the Missouri Botanical Garden* **61**, 539-673.

RECHER, H.F. (1986). Integrating national park and forestry objectives for wildlife conservation: The forester's responsibility. *The Forestry Log* **18**, 5-7.

RECHER, H.F., and CHRISTENSEN, P.E. (1981). Fire and the evolution of the Australian biota. In *Ecological Biogeography of Australia*. Ed. A. Keast. Junk, The Hague. pp. 135-162.

RECHER, H.F., SHIELDS, J., KAVANAGH, R. and WEBB, G. (1987). Retaining remnant mature forest for nature conservation at Eden, New South Wales: A review of theory and practice. In: *Nature Conservation: The Role of Remnants of Native Vegetation*. Eds. D.A. Saunders, G.W. Arnold, A.A. Burbidge and A.J.M. Hopkins. Surrey Beatty and Sons, Chipping Norton, N.S.W. pp. 177-194

RIEK, E.F. (1967). The freshwater crayfish of Western Australia (Decapoda: Parastacidae). *Australian Journal of Zoology* **15**, 103-121.

ROBERTS, J.D., WARDELL-JOHNSON, G. and BARENDSE, W. (1990). Extended descriptions of *Geocrinia vitellina* and *Geocrinia alba* (Anura:Myobatrachidae) from south-western Australia, with comments on the status of *G. lutea*. *Records of the Western Australian Museum* **144** (4), 427-437.

ROGNON, P., and WILLIAMS, M.A.J. (1977). Late Quaternary climatic changes in Australia and North Africa: a preliminary interpretation. *Palaeogeography, Palaeoclimatology, Palaeoecology* **21**, 285-327.

ROSEN, D.E. (1974). Phylogeny and zoogeography of salmoniform fishes and relationships of *Lepidogalaxias salamandroides*. *Bulletin of American Museum of Natural History* **153**, 265-325.

ROTHERAM. I. (1982). Suppression of growth surrounding regeneration by veteran trees of karri (*Eucalyptus diversicolor*). *Australian Forestry* **46** (1), 8-13.

ROURKE, J. and WIENS, D. (1977). Convergent floral evolution in South African and Australian Proteaceae and its possible bearing on pollination by non-flying mammals. *Annals of the Missouri Botanical Gardens* **64**, 1-17.

SAMMY, N. (1983). Distribution infomation for Western Australian lichens. Western Australian Herbarium Report.

SAMMY, N. (1985). The distribution of lichens in Western Australia. *W.A. Herbarium Research Notes* **11**, 53 -113.

SAMPSON, J.C. (1971). The Biology of *Bettongia penicillata* Gray, 1837. PhD Thesis, University of W.A., Nedlands.

SAYER, F.W. (1969). *Western Australia Yearbook.* Commonwealth Bureau of Census and Statistics,. Government Printer W.A.

SCHOFIELD, N.J., RUPRECHT, J.K. and LOH, I.C. (1988). The impact of agricultural development on the salinity of surface water resources of south-west Western Australia. Water Authority of W.A. Report No. WS27.

SCHUSTER, R.M. (1972). Continental movements, "Wallaces Line" and Indomalayan-Australasian dispersal of land plants: some eclectic concepts. *Botanical Review* **38,** 3-86.

SCHUSTER, R.M. (1976). Plate tectonics and its bearing on the geographical origin and dispersal of angiosperms. In *Origin and early evolution of angiosperms*. Ed. C.B. Beck. Columbia University Press, London. pp. 48-138.

SCOTT, D.R.M. (1970). Natural plant communities: concepts and misconceptions. In: Natural areas - needs and opportunities. Proceedings of a symposium. Eds W.J. Dittrich and J.M. Trappe. Continuing Education Publications, Corvallis, Oregon. pp. 15-16.

SCOTT, G.A.M., STONE, I.G. and ROSSER, C. (1976). *The Mosses of Southern Australia.* Academic Press. London.

SEDDON, G. (1972). *A Sense of Place.* University of W.A. Press, Nedlands.

SERENA, M., SODEQUIST, T.R. and MORRIS, K. (1991). *The Chuditch.* Department of Conservation and Land Management (W.A.). Wildlife Management Program No. 5.

SERVENTY, D.L. and WHITTELL, H.M. (1976). *Birds of Western Australia.* 5th edn. University of W.A. Press, Nedlands.

SERVENTY, V.N. (1952). Cave paintings near York and Hyden. *Western Australian Naturalist* **3,** 121-130.

SHEARER, B.L. and BATINI, F.E. (1990). *Phytophthora* and other fungal diseases. In: The Hill River Project. Eds. A.A. Burbidge and S. van Leeuwin. Department of Conservation and Land Management (W.A.) Occasional Paper 1/90.

SHEARER, B.L. and TIPPETT, J.T. (1989). Jarrah Dieback: the dynamics and management of *Phytophthora cinnamomi* in the jarrah *(Eucalyptus marginata)* forest of South-Western Australia. Department of Conservation and Land Management, Research Bulletin No. 3.

SIMON, E. (1909) Araneae, 2me Partie. *Die Fauna Sudwest-Australiens* **2**(12), 155-212, (Jena).

SLATER, P. (1974). *A Field Guide to Australian Birds, Passerines.* Rigby, Adelaide.

SMITH, F.G. (1972). *Vegetation Survey of Western Australia, Vegetation Map: Pemberton (and) Irwin Inlet.* Department of Agriculture, Perth, W.A.

SMITH, G.S. (1973). *Vegetation Survey of Western Australia, Vegetation Maps of Busselton and Augusta.* Department of Agriculture, Perth, W.A.

SMITH, R. (1951a). *Soils of the Margaret River - Lower Blackwood River Districts, Western Australia.* CSIRO Bulletin 262.

SMITH, R. (1951b). *Pedogenesis in the Frankland River Valley, Western Australia.* CSIRO Bulletin 265.

SOLEM, A, GIRARDI, E.L, SLACK-SMITH, S, and KENDRICK, G.W. (1982). *Austroassiminea letha*, gen. nov., sp. nov., a rare and endangered prosobranch snail from south-western Australia (Mollusca: Prosobranchia: assimineidae). *Journal of the Royal Society of Western Australia* **65** (4), 119-129.

SPECHT, H.I. and CLELLAND, J.B. (1961). Flora conservation in South Australia. I. The preservation of plant formations and associations recorded in South Australia. *Transactions of the Royal Society of South Australia* **84,** 177-197.

SPECHT, R.H., POE, E.M. and BOUGHTON, V.H. (1974). Conservation of major plant communities in Australia and Papua New Guinea. *Australian Journal of Botany* Supplementary Series **7**, 1-667.

SPECHT, R.L. (1969). A comparison of the sclerophyllous vegetation characteristic of Mediterranean-type climates in France, California, and southern Australia. I. Structure, morphology and succession. II. Dry matter, energy and nutrient accumulation. *Australian Journal of Botany* **17,** 277-292, 293-308.

SPECHT, R.L. (1970). Vegetation. In *The Australian Environment.* Ed. G.W. Leeper. CSRIO and Melbourne University Press, Melbourne. pp. 44-76.

SPECIAL REVIEW COMMITTEE (1976). A review of recommendations for reserves in the south-west and south coastal areas of Western Australia. Environmental Protection Authority (W.A.) Perth.

SPECK, N.H. (1958). The vegetation of the Darling-Irwin botanical districts and an investigation of the distribution patterns of the family Proteaceae, south-western Australia. Ph.D Thesis. Department of Botany, University of W.A., Nedlands.

START, A.N. (1982). Report on 6 months search for potoroos in Western Australia. Department of Fisheries and Wildlife (W.A.). Report.

STEPHENS, C.G. (1946). *Pedogenesis following dissection of lateritic regions in southern Australia.* CSIRO Bulletin No 206.

STOATE, T.N. and KESSELL, S.L. (1926). Working Plan No. 1: Mundaring Working Circle. Forests Department (W.A.) Bulletin 37.

STONEMAN, G.L., ROSE, P.W., and BORG, H. (1988). *Recovery of Forest Density after Intensive logging in the Southern Forest of Western Australia.* Department of Conservation and Land Management (W.A.) Technical Report No 19.

STOUTAMIRE, W.P. (1983). Wasp pollinated species of *Caladenia* (Orchidaceae) in south-western Australia. *Australian Journal of Botany* **31,** 383-394.

STRAHAN, R. (1983). *The Australian Museum Complete Book of Australian Mammals.* The National Photographical index of Australian Widlife, Angus and Robertson, Sydney.

STRELEIN, G.J. (1988). Site Classification in the Southern Jarrah Forest of Western Australia. Department of Conservation and Land Management (W.A.) Research Bulletin No 2.

STUART-STREET, A. and KIRKPATRICK, B. (1991) Landscape character types of Western Australia. A preliminary draft. Unpublished report, Department of Conservation and Land Management, Como, W.A.

STUART-STREET, A. and KIRKPATRICK, B. (in prep). Visual landscape character types of Western Australia.

THEISCHINGER, G. and WATSON, J.A.L. (1984). Larvae of Australian Gomphomacromiinae and their bearing on the status of the Synthemis group of genera, Odonata Corduliidae. *Australian Journal of Zoology* **32,** 67-95.

TINDALE, N.B. (1974). *Aboriginal Tribes of Australia.* University of California Press, Los Angeles.

TINGAY, A. and TINGAY, S.R. (1984). *Bird communities in the karri forest of Western Australia.* Australian Conservation Foundation (Inc), Hawthorn, Vic.

TYNDALE-BISCOE, C.H. and CALABY, J.H. (1975). Eucalypt forest as refuge for wildlife. *Australian Forestry* **38,** 117-133.

UNDERWOOD, R.J. (1978). Natural fire periodicity in the karri (*Eucalypt diversicolor* F. Muell) Forest. Forests Department (W.A.) Research Paper 41.

UNDERWOOD, R.J. and CHRISTENSEN, P.E.S. (1981). Forest fire management in Western Australia. Forest Department (W.A.) *Special Focus No. 1,*

UNDERWOOD, R.J. and MURCH, J.H. (1983). Hygienic logging in the northern jarrah forest. *Australian Forestry,* **47**(1), 39-44.

UNDERWOOD, R.J., SNEEUWJAGT, R.J. and STYLES, H.G. (1985). The contribution of prescribed burning to forest fire control in Western Australia: Case studies. In *Fire Ecology and Management in Western Australia ecosystems* . Ed. J.R. Ford. WAIT Environmental Studies Group, Report No. 14. Western Australian Institute of Technology, Perth. pp. 153-170.

WALKER, D. and SINGH, G. (1981). Vegetation History. In *Australian Vegetation*. Ed. R.H. Groves. Cambridge University Press, Cambridge. pp. 26-43.

WARDELL-JOHNSON, G.W. (1985). The composition and foraging ecology of a bird community in karri forest in southwestern Australia. MSc. Thesis, Oxford University.

WARDELL-JOHNSON, G.W. (1986). The use of nest boxes by Mardos (Antechinus flavipes leucogaster) in regenerating karri forest in South Western Australia. *Australian Wildlife Research* 13, 407-417.

WARDELL-JOHNSON, G. and CHRISTENSEN, P. (1992). A review of the effects of disturbance on wildlife of the karri forest. In: Research on the impact of forest management in south-west Western Australia. Department of Conservation and Land Management (W.A.) Occasional Paper 2/92.

WARDELL-JOHNSON, G.W. and ROBERTS, D. (1991a). The survival status of the *Geocrinia rosea* (Anura:Myobatrachidae) complex in riparian corridors: biogeographical implications, In: *Nature Conservation. The role of corridors*. Eds. D. Saunders and R.J. Hobbs. Surrey Beatty and Sons, Chipping Norton, N.S.W. pp. 167 - 175

WARDELL-JOHNSON, G. and ROBERTS, D. (1991b). Endangered forest frogs. In *Landscope*. Department of Conservation and Land Management, W.A.

WARDELL-JOHNSON, G., INIONS, G. and ANNELS, A. (1989). A floristic classification of the Walpole-Nornalup National Park, Western Australia. *Forest Ecology and Management* 28, 259-279.

WATSON, J.R. and O'FARRELL, A.F. (1985). The affinities of the Australian Gomphidae (Odonata). *Australian Journal of Zoology* 33, 499-508.

WEBB, L.J. (1966). The identification and conservation of habitat types in the wet tropical lowlands of North Queensland. *Proceedings of the Royal Society of Queensland* 78, 59-86.

WEBB, L.J. (1970). Fire environments in eastern Australia. *Second Fire Ecology Symposium*. Monash University, Vic. 1970 pp. 1-5.

WHITE, B.J. (1974a). *Clear felling with seed trees*. Forests Department (W.A.) Research Paper 13.

WHITE, B.J. (1974b) *Karri Silvics*. Forests Department, W.A.

WHITE, J.P. and O'CONNELL, J.F. (1982). *A Prehistory of Australia, New Guinea and Sahul*. Academic Press, Sydney,.

WHITE, M.J.D. (1978). *Modes of Speciation*. Freeman, San Francisco.

WHITELY, G.P. (1947). The fluvifaunae of Australia with particular reference to freshwater fishes in Western Australia. *Western Australian Naturalist* 1, 49-53.

WILDE, S.A. and WALKER, I.W. (1984). Pemberton-Irwin Inlet, Western Australia. W.A. Geological Survey 1:250 000. In Geological Series Explanatory Notes.

WILDLIFE CONSERVATION ACT (1950-1979). No 77 of Western Australia, Government Printer W.A.

WILLIAMS, W.D. (1981). The Crustacea of Australian inland waters. In *Ecological Biogeography of Australia*. Ed. A. Keast. Junk, The Hague. pp. 1101-1138.

WOOLLER, R.D., RUSSELL, E.M. and RENFREE, M.B. (1984). Honey possums and their food plants. In *Possums and gliders*. Eds. A.P. Smith and I.R. Hume. Surrey Beatty, Chipping Norton, N.S.W.

WORSLEY ALUMINA (1985). Worsley Alumina Project: flora and fauna studies Phase two Worsley Alumina Pty Ltd, Perth, W.A.

WRONSKI, E.B. (1984). Impact of tractor thinning operations on soil and tree roots in a karri forest, Western Australia. *Australian Forestry Research* 14, 319-332.

APPENDICES

Appendix 1
SOUTHERN FORESTS FLORA

This list comprises the species listed for the Warren Botanical Subdistrict by Hopper *et al.* (1992)[a] together with additional species from outside the Warren Botanical Subdistrict which occur within the southern forests from Christensen *et al* (1985)[b].
Nomenclature and systematic arrangement follow Green (1985)[c].

FERNS AND FERN ALLIES

LYCOPODIACEAE
Lycopodium serpentinum Kunze
Phylloglossum drummondii Kunze

SELAGINELLACEAE
Selaginella gracillima (Kunze) Alston

ISOETACEAE
Isoetes australis S. Williams
Isoetes drummondii A. Braun
Isoetes sp.

OPHIOGLOSSACEAE
Ophioglossum lusitanicum L. ssp. *coriaceum* (A. Cunn.) Clausen

SCHIZAEACEAE
Schizaea fistulosa Labill.

ADIANTACEAE
Adiantum aethiopicum L.
Anogramma leptophylla (L.) Link
Cheilanthes austrotenuifolia Quirk et Chambers

PTERIDACEAE
Pteris vittata L.

CYATHACEAE
* *Sphaeropteris cooperi* (Hook ex F. Muell.) Domin

DENNSTAEDTIACEAE
* *Hypolepis rugulosa* (Labill.) J. Smith
Pteridium esculentum (G. Forster) Cockayne

LINDSAEACEAE
Lindsaea linearis Sw.

ASPLENIACEAE
Asplenium aethiopicum (Burm. f.) Bech.
Asplenium flabellifolium Cav.
Asplenium obtusatum G. Forster

MARSILEACEAE
Pilularia novae-hollandiae A. Braun

GYMNOSPERMS

ZAMIACEAE
Macrozamia riedlei (Fischer. ex Gaudich.) C. Gardner.

PODOCARPACEAE
Podocarpus drouynianus F. Muell.

---

* = Introduced
(E) = endemic to the Warren Botanical Subdistrict

[a]HOPPER, S.D., KEIGHERY G.J. and WARDELL-JOHNSON (1992). Flora of the karri forest and other communities in the Waren Botanical Subdistrict of Western Australia. In: Research on the impact of forest management in south-west Australia. Department of Conservation and Land Management Occasional Paper 2/92.

[b] CHRISTENSEN, P., ANNELS, A., LIDDELOW, G. and SKINNER, P. (1985). Vertebrate fauna in the southern forests of Western Australia: a survey. Forests Department (W.A.) Bulletin 94.

[c] GREEN, J.W. (1985). *Census of the vascular plants of Western Australia.* Department of Agriculture (W.A.) Perth.

PINACEAE

* *Pinus pinaster* Aiton
* *Pinus radiata* D. Don

CUPRESSACEAE

Actinostrobus pyramidalis Miq.

MONOCOTYLEDONS

TYPHACEAE

* *Typha orientalis* C. Presl

RUPPIACEAE

Ruppia megacarpa Mason
Ruppia polycarpa Mason

POTAMOGETONACEAE

Potamogeton drummondii Benth.
Potamogeton ochreatus Raoul

ZANICHELLIACEAE

Lepilaena bilocularis Kirk
Lepilaena cylindrocarpa (Koern. ex Walp.) Benth.
Lepilaena preissii (Lehm.) F. Muell.

POSIDONIACEAE

Posidonia australis J.D. Hook. f.
Posidonia denhartogii Kuo et Cambridge
Posidonia kirkmanii Kuo et Cambridge
Posidonia ostenfeldii Hartog
Posidonia robertsoniae Kuo et Cambridge
Posidonia sinuosa Cambridge et Kuo

CYMODOCEACEAE

Amphibolus antarctica (Labill.) Sonder
Amphibolus griffithii (Black) Hartog
Thalassodendron pachyrhizum Hartog

ZOSTERACEAE

Heterozostera tasmanica (Martens ex Asch.) Hartog

NAJADACEAE

Najas marina L.

JUNCAGINACEAE

Triglochin calcitrapa Hook.
Triglochin centrocarpa Hook.
Triglochin minutissima F. Muell.
Triglochin procera R. Br.
Triglochin striata Ruiz. et Pav

HYDROCHARITACEAE

Halophila ovalis (R. Br.) J.D. Hook.
Ottelia ovalifolia (R. Br.) Rich.
* *Vallisneria spiralis* L.

POACEAE

Agrostis avenacea J. Gmolin
Agrostis drummondiana (Steud.) Vickery
* *Agrostis gigantea* Roth
* *Agrostis stolonifera* L.
Agrostis venusta Trin.
* *Aira caryophyllea* L.
* *Aira cupaniana* Guss.
* *Alopecurus myosuroides* Hudson
* *Ammophila arenaria* (L.) Link
Amphipogon amphipogonoides (Steud.) Vick.
Amphipogon debilis R. Br.
Amphipogon laguroides R. Br.
Amphipogon turbinatus R. Br.
* *Anthoxanthum odoratum* L.
* *Arrhenatherum bulbosum* (Willd.) C. Presl
* *Arundo donax* L.
* *Avellinia michelii* (Savi) Parl.
* *Avena barbata* Link
* *Axonopus compressus* (Swartz.) Beauv.
* *Briza maxima* L.
* *Briza minor* L.
Bromus arenarius Labill.
* *Bromus catharticus* M. Vahl.
* *Bromus hordeaceus* L.
* *Cortaderia selloana* (Schultes et J.H. Schultes) Asch. et Graebn.
* *Cynodon dactylon* (L.) Pers.
* *Cynosurus cristatus* L.
* *Cynosurus echinatus* (L.)
* *Dactylis glomerata* L.
Danthonia caespitosa Gaud.
Danthonia pilosa R. Br.
Danthonia racemosa R. Br.
Danthonia setacea R. Br. var. *setacea*
Deyeuxia quadriseta Benth.
Dichelachne crinita (L.f.) J.D. Hook.
Dichelachne micrantha (Cav.) Domin
* *Digitaria sanguinalis* (L.) Scop.
Diplopogon setaceus R. Br.
* *Echinochloa crusgalli* (L.) Beauv.
* *Echinopogon ovatus* (G. Forster) P. Beauv.
* *Ehrharta erecta* Lam.
Erharta villosa J.H. Schultes
* *Ehrharta calycina* Smith
* *Ehrharta longiflora* Smith
* *Ehrharta pusilla* Nees ex Trin.
Eragrostis benthamii Mattei
* *Eragrostis curvula* (Schrad.) Nees
Eragrostis elongata (Willd.) Jacq.
* *Festuca arundinacea* Schreber
Festuca littoralis Labill.
Glyceria australis C.E. Hubb.
* *Glyceria maxima* (Hartm.) O.R. Holmberg
* *Hainardia cylindrica* (Willd.) W. Greuter
Hemarthria uncinata R. Br.
* *Holcus lanatus* L.
* *Hordeum glaucum* Steudel
* *Lagurus ovatus* L.
* *Lolium perenne* L.
* *Lolium rigidum* Gaudin
* *Lolium temulentum* L.
* *Melinis minutiflora* P. Beauv.
Microlaena stipoides (Labill.) R. Br.
Neurachne alopecuroidea R. Br.
* *Panicum capillare* L.

* *Panicum maximum* Jacq.
* *Parapholis incurva* (L.) C.E. Hubb.
Paspalum dilatatum Poit
* *Paspalum distichum* L.
* *Paspalum vaginatum* Sw.
Pennisetum clandestinum Hochst. f. ex Chiov.
* *Phleum pratense* L.
* *Phragmites australis* (Cav.) Trin. ex Steud.
* *Piptatherum miliaceum* (L.) Cosson
* *Poa annua* L.
Poa drummondiana Nees
Poa homomalla Nees
Poa poiformis (Labill.) Druce
Poa porphyroclados Nees
* *Poa pratensis* L.
Poa serpentum Nees
* *Polypogon monspeliensis* (L.) Desf.
Polypogon tenellus R. Br.
* *Polypogon viridis* (Gouan) Breistr.
* *Sorghum halepense* (L.) Pers.
Spinifex hirsutus Labill.
* *Sporobolus indicus* (L.) R. Br.
Sporobolus virginicus (L.) Kunth
* *Stenotaphrum secundatum* (Walter) Kunze
Stipa campylachne Nees
Stipa compressa R. Br.
Stipa flavescens Labill.
Stipa hemipogon Benth.
Stipa macalpinei Reader
Stipa mollis R. Br.
Stipa semibarbata R. Br.
Stipa tenuifolia Steud.
Stipa trichophylla Benth.
Tetrarrhena laevis R. Br.
* *Vulpia membranacea* (L.) Dumort
* *Vulpia myuros* Gmel.

CYPERACEAE
Baumea acuta (Labill.) Palla
Baumea articulata (R. Br.) S.T. Blake
Baumea juncea (R.Br.) Palla
Baumea rubiginosa (Sprengel) Boeck.
Bolboschoenus caldwellii (Cook) Sojak
Carex appressa R. Br.
* *Carex divisa* Hudson
Carex fascicularis Sol. ex Boott
Carex inversa R. Br.
Chorizandra cymbaria R. Br.
Chorizandra enodis Nees
Cyathochaeta avenacea Benth.
Cyathochaeta clandestina (R. Br.) Benth.
* *Cyperus congestus* Vahl
Cyperus laevigatus L.
* *Cyperus tenellus* L.f.
Evandra aristata R. Br.
Evandra pauciflora R. Br.
Fimbristylis velata R. Br.
Gahnia decomposita (R.Br.)Benth.
Gahnia deusta (R. Br.) Benth.
Gahnia filum (Labill) F. Muell.
Gahnia trifida Labill.
(E) *Gymnoschoenus anceps* (R. Br.) C.B. Clarke
Isolepis cernua (M. Vahl.) Roemer et Schultes
Isolepis congrua Nees
Isolepis cyperoides R. Br.
Isolepis fluitans (L.) R. Br.
Isolepis marginata (Thunb.) A. Dietr.
Isolepis nodosa (Rottb.) R. Br.
Isolepis oldfieldiana (S.T. Blake) K.L. Wilson
* *Isolepis prolifer* (Rottb.) R. Br.
Isolepis stellata (C.B. Clarke) K.L. Wilson
Lepidosperma angustatum R. Br.
Lepidosperma brunonianum Nees
Lepidosperma effusum Benth.
L epidosperma gladiatum Labill.
Lepidosperma gracile R. Br.
Lepidosperma leptophyllum Benth.
Lepidosperma leptostachyum Benth.
Lepidosperma longitudinale Labill.
(E) *Lepidosperma persecans* S.T. Blake
Lepidosperma squamatum Labill.
Lepidosperma tenue Benth.
Lepidosperma tetraquetrum Nees
Mesomelaena graciliceps (C.B. Clarke) K.L. Wilson
Mesomelaena stygia (R. Br.) Nees
Mesomelaena tetragona (R. Br.) Benth.
Reedia spathacea F. Muell.
Schoenoplectus validus (M. Vahl.) A. Love et D. Love
(?E) *Schoenus acuminatus* R. Br.
Schoenus asperocarpus F. Muell.
Schoenus bifidus (Nees) Boeckler
Schoenus breviculmis Benth.
Schoenus brevisetis (R. Br) Benth
Schoenus caespititius W. Fitzg.
(E) *Schoenus cruentus* (Nees) Benth.
Schoenus curvifolius (R. Br.) Benth.
(E) *Schoenus efoliatus* F. Muell.
Schoenus grammatophyllus F. Muell
Schoenus grandiflorus (Nees) F. Muell.
Shoenus lanatus Labill
(?E) *Schoenus multiglumis* Benth.
Schoenus nitens (R. Br.) Poiret
Schoenus odontocarpus F. Muell.
Schoenus pleistomoneus F. Muell
Schoenus rodwayanus W. Fitzg.
Schoenus subbulbosus Benth.
Schoenus sublaxus Kuek.
Schoenus trachycarpus F. Muell.
Tetraria capillaris (F. Muell.) J. Black
Tetraria octandra (Nees) Kuek.
Tricostularia neesii Lehm. var. *neesii*
Tricostularia neesii var. *elatior* Benth.

ARACEAE
* *Zantedeschia aethiopica* (L.) Sprengel

LEMNACEAE
Lemna disperma Hegelm.

RESTIONACEAE
(E) *Alexgeorgea ganopoda* L. Johnson et Briggs
Anarthria gracilis R. Br.
Anarthria laevis R. Br.

Anarthria prolifera R. Br.
Anarthria scabra R. Br.
Chaetanthus leptocarpoides R. Br.
Ecdeiocolea monostachya F. Muell.
Empodisma gracillimum (F. Muell.) L. Johnson et Cutler
Hypolaena exsulca R. Br.
Hypolaena ramosissima Gilg.
Hypolaena sp.
Lepidobolus sp. aff. *chaetocephalus* (? *preissianus* Nees)
Leptocarpus aristatus R. Br.
Leptocarpus canus Lindley et Nees
Leptocarpus coangustatus Nees
Leptocarpus scariosus R. Br.
Leptocarpus tenax (Labill.) R. Br.
Leptocarpus tenellus (Nees) F. Muell.
Leptocarpus sp. "d"
Leptocarpus sp. "r"
(E) *Leptocarpus* sp. "t"
Lepyrodia drummondiana Steudel
Lepyrodia glauca (Nees) F. Muell.
Lepyrodia heleocharoides Gilg.
Lepyrodia hermaphrodita R. Br.
Lepyrodia muirii F. Muell.
Lepyrodia stricta R. Br.
Loxocarya cinerea R. Br.
Loxocarya fasciculata (R. Br.) Benth.
Loxocarya flexuosa (R. Br.) Benth.
Loxocarya pubescens (R. Br.) Benth.
(E) *Loxocarya* sp.
Lopxocarya aff *cinerca* "r"
Lyginia barbata R. Br.
Meeboldina denmarkica Suesseng
"Pseudoloxocarya" grossa (gen/sp. nov.)
Restio amblycoleus F. Muell
Restio applanatus Sprengel
Restio "crassus"
Restio gracilior (F. Muell) Benth.
Restio laxus R. Br.
Restio leptocarpoides Benth.
Restio tremulus R. Br.
Restio ustulatus F. Muell ex Ewart et Sharman

CENTROLEPIDACEAE
Aphelia cyperoides R. Br.
Aphelia drummondii (Hieron.) Benth.
Centrolepis aristata (R. Br.) Roem. et Schult.
(E) *Centrolepis caespitosa* D.A. Cooke
Centrolepis drummondiana (Nees) Walp.
Centrolepis fascicularis Labill.
Centrolepis glabra (F. Muell. ex Sond.) Hieron
Centrolepis humillima F. Muell. ex Benth.
Centrolepis inconspicua W. Fitzg.
Centrolepis pilosa Hieron
Centrolepis mutica (R. Br.) Hieron
Centrolepis polygyna (R. Br.) Hieron
Centrolepis strigosa (R. Br.) Roem et Schult.

HYDATELLACEAE
Trithuria bibracteata D.A. Cooke
Trithuria submersa J.D. Hook.

XYRIDACEAE
Xyris flexifolia R. Br.
Xyris gracillima F. Muell.
Xyris indivisa Wakef.
Xyris lacera R. Br.
Xyris lanata R. Br.
Xyris laxiflora F. Muell.
(E) *Xyris roycei* Wakef.

PHILYDRACEAE
Philydrella drummondii L.G. Adams
Philydrella pygmaea (R. Br.) Caruel

JUNCACEAE
Juncus amabilis E. Edgar
* *Juncus articulatus* L.
* *Juncus bufonius* L.
Juncus caespiticius E. Meyer
* *Juncus capitatus* Weigel
Juncus gregiflorus L. Johnson
Juncus holoschoenus R. Br.
Juncus kraussii Hochst.
* *Juncus microcephalus* Kunth
* *Juncus oxycarpus* E. Meyer
Juncus pallidus R. Br.
Juncus pauciflorus R. Br.
Juncus planifolius R. Br.
Juncus prismatocarpus R. Br.
Juncus procerus E. Meyer
Juncus subsecundus Wakef.
Luzula meridionalis Nordensk.

ASPARAGACEAE
* *Myrsiphyllum asparagoides* (L.) Willd.

DASYPOGONACEAE
Acanthocarpus preissii Lehm.
Baxteria australis R. Br.
Calectasia grandiflora Preiss
Chamaexeros serra (Endl.) Benth.
(E) *Chamaexeros* sp. nov.
Dasypogon bromeliifolius R. Br.
Dasypogon hookeri I. Drumm.
Kingia australis R. Br.
Lomandra brittanii T.S. Choo
Lomandra caespitosa (Benth.) Ewart
Lomandra drummondii (F. Muell. ex Benth.) Ewart
Lomandra hastilis (R. Br.) Ewart
Lomandra hermaphrodita (C.R.P. Andrews) C.A. Gardner
Lomandra integra T.D. Macfarlane
Lomandra micrantha (Endl.) Ewart ssp. *micrantha*
Lomandra nigricans MacFarlane
Lomandra odora (Endl.) Ewart
(E) *Lomandra ordii* F. Muell.
Lomandra pauciflora (R. Br.) Ewart
Lomandra preissii (Endl.) Ewart
Lomandra purpurea (Endl.) Ewart
Lomandra sericea (Endl.) Ewart
Lomandra sonderi (F. Muell.) Ewart
Lomandra suaveolens (Endl.) Ewart

XANTHORRHOEACEAE
Xanthorrhoea gracilis Endl.
Xanthorrhoea preissii Endl.
Xanthorrhoea platyphylla D.J. Bedford

PHORMIACEAE
Dianella divaricata R. Br.
Dianella revoluta R. Br. var. *brevicaulis* Ostenf.
Stypandra grandiflora Lindley

ANTHERICACEAE
Agrostocrinum sp.
Agrostocrinum scabrum (R. Br.) Baillon
Arthropodium preissii Lehm. ex Endl.
Borya constricta D.M. Churchill
Borya longiscapa D.M. Churchill
Borya nitida Labill.
Borya seirboidea Lindley
Borya sphaerocephala R. Br.
Caesia micrantha Lindley
Caesia occidentalis R. Br.
Chamaescilla corymbosa R. Br. var. *corymbosa*
Chamaescilla corymbosa var. *latifolia* (F. Muell.) R. Henderson
Chamaescilla spiralis (Endl.) F. Muell.
Corynotheca micrantha (Lindley) Macbride var. *panda* Henderson
Hodgsoniola junciformis (F. Muell.) F. Muell.
Johnsonia acaulis Endl.
Johnsonia inconspicua G.J. Keighery
Johnsonia lupulina R. Br.
Johnsonia teretifolia Endl.
Laxmannia jamesii G.J. Keighery
Laxmannia minor R. Br.
Laxmannia ramosa Lindley
Laxmannia sessiliflora Deene
Sowerbaea laxiflora Lindley
Thysanotus arbuscula Baker
Thysanotus dichotomus (Labill.) R. Br.
Thysanothus arenarius N.H. Brittain
(E) *Thysanotus formosus* N.H. Brittan
Thysanotus gracilis R. Br.
(E) *Thysanotus isantherus* R. Br.
Thysanotus manglesianus Kunth
Thysanotus multiflorus R. Br.
Thysanotus patersonii R. Br.
Thysanotus pauciflorus R. Br.
Thysanotus pseudojunceus N.H. Brittan
Thysanotus sparteus R. Br.
Thysanotus tenellus Endl.
Thysanotus thyrsoideus Baker
Thysanotus triandrus (Labill.) R. Br.
Tricoryne elatior R. Br.
Tricoryne humilis Endl.

ASPHODELACEAE
Bulbine semibarbata (R. Br.) Haw.
* *Trachyandra divaricata* (Jacq.) Kunth

HYACINTHACEAE
* *Albuca canadensis* (L.) F.M. Leighton

ALLIACEAE
* *Allium ampeloprasum* L.
* *Allium triquetrum* L.
* *Nothoscordum gracile* (Aiton) Stearn

COLCHICACEAE
Burchardia monantha Domin
Burchardia multiflora Lindley
Burchardia umbellata R. Br.
Wurmbea dioica (R. Br.) F. Muell. ssp. alba Macfarlane
Wurmbea monantha (Endl.) T.D. Macfarlane
Wurmbea sp.

HAEMODORACEAE
Anigozanthos bicolor Endl. ssp. *decrescens* Hopper
Anigozanthos flavidus Redoute et DC.
Anigozanthos manglesii D. Don ssp. *manglesii*
Anigozanthos preissii Endl.
Anigozanthos viridis Endl. ssp. *viridis*
Conostylis aculeata R. Br. ssp. *aculeata*
Conostylis aculeata ssp. *gracilis* (ined.).
Conostylis candicans Endl. ssp. *calcicola* Hopper
Constylis laxiflora Benth.
Conostylis serrulata R. Br.
Conostylis setigera R. Br. ssp. *setigera*
Conostylis setosa Lindley
Haemodorum discolor T.D. Macfarlane
Haemodorum laxum R. Br.
Haemodorum paniculatum Lindley
Haemodorum simplex Lindley
Haemodorum sparsiflorum F. Muell.
Haemodorum spicatum R. Br.
Phlebocarya ciliata R. Br.
Tribonanthes australis Endl.
Tribonanthes brachypetala Lindley
Tribonanthes violacea Endl.

AMARYLLIDACEAE
* *Agapanthus praecox* Willd. ssp. *praecox*
* *Amaryllis belladonna* L.
* *Narcissus tazetta* L.

HYPOXIDACEAE
Hypoxis glabella R. Br. var. *glabella*
Hypoxis occidentalis Benth. var. *quadriloba* Henderson

IRIDACEAE
* *Babiana stricta* (Ait.) Ker-Gawler
* *Chasmanthe floribunda* (Salisb.) N.E. Br.
* *Crocosmia x crocosmiiflora* (Lemoine) N.E. Br.
* *Ferraria crispa* Burman
* *Freesia leichtlinii* Klatt
* *Gladiolus angustus* L.
* *Gladiolus carneus* Del.
* *Gladiolus undulatus* L.
* *Gynandriris setifolia* (L.f.) R. Foster
* *Homeria flaccida* Sweet
* *Homeria miniata* (Andr.) Sweet
* *Iris germanica* L.
* *Iris unguicularis* Poiret
* *Ixia maculata* L.

* *Ixia paniculata* Delaroche
Orthrosanthus laxus (Endl.) Benth. var. *laxus*
Orthrosanthus laxus var. *gramineus* (Endl.) Geer.
Orthrosanthus multiflorus Sweet
(E) *Orthrosanthus polystachyus* Benth.
Patersonia babianoides Benth.
Patersonia juncea Lindley
Patersonia limbata Endl.
Patersonia occidentalis R. Br.
Patersonia pygmaea Lindley
Patersonia umbrosa Endl. var. *umbrosa*
Patersonia umbrosa var. *xanthina* (F.Muell.)Geer.
* *Romulea rosea* (L.) Ecklon
* *Sparaxis bulbifera* (L.) Ker Gawler
* *Tritonia lineata* (Salisb.) Ker Gawler
* *Watsonia bulbillifera* J. Mathews
* *Watsonia leipoldtii* L. Bolus
* *Watsonia marginata* (L.f.) Ker Gawler
* *Watsonia versfeldii* J. Mathews et L. Bolus
var. *alba* J. Mathews et L. Bolus
* *Watsonia wordsworthiana* J. Mathews et L. Bolus

ORCHIDACEAE
(E) *Caladenia abbreviata* Hopper et A.P. Brown.
(E) *Caladenia* aff. *abbreviata* Hopper et A.P. Brown.
Caladenia applanata Hopper A.P. Brown
ssp. *applanata*
(E) *Caladenia applanata* ssp. *erubescens*
Hopper et A.P. Brown.
Caladenia arrecta Hopper et A.P. Brown.
Caladenia attingens Hopper ined. ssp. *attingens*
Caladenia barbarossa H.G. Reichb.
Caladenia bicalliata Rogers
Caladenia brownii Hopper et A.P. Brown.
Caladenia busselliana Hopper et A.P. Brown.
Caladenia caesarea Clements et Hopper
ssp. *maritima* Hopper ined.
Caladenia cairnsiana F. Muell.
Caladenia chapmanii Hopper et A.P. Brown
ssp. *chapmanii*
Caladenia chapmanii ssp. *variegata* Hopper ined.
Caladenia citrina Hopper et A.P. Brown.
Caladenia corynephora A.S. George
Caladenia dilatata R. Br. var *falcata*
Caladenia discoidea Lindley
Caladenia evanescens Hopper et A.P. Brown
Caladenia filamentosa R. Br.
Caladenia ensata Nicholls
Caladenia excelsa Hopper et A.P. Brown.
Caladenia ferruginea Nicholls
Caladenia flava R. Br. ssp. *flava*
Caladenia flava ssp. *sylvestris* Hopper et A.P. Brown
(E) *Caladenia gardneri* Hopper et A.P. Brown
Caladenia gardineri ssp. *valida* (in edit).
Caladenia grandis (ined.).
(E) *Caladenia harringtoniae* Hopper et A.P. Brown
Caladenia heberleana Hopper et A.P. Brown
Caladenia hirta Lindley ssp. *hirta*
(E) *Caladenia huegelii* H.G. Reichb. ssp. *redacta*
Hopper et A.P. Brown
Caladenia hummiliflora Hopper et A.P. Brown
ssp. *meridionalis* Hopper et A.P. Brown.
Caladenia infundibularis A.S. George
Caladenia interjacens Hopper et A.P. Brown
Caladenia latifolia R. Br.
Caladenia longicauda Lindley ssp. *longicauda*
Caladenia longicauda Lindley
ssp. *splendens* Hopper et A.P. Brown
Caladenia longiclavata E. Coleman var. *longiclavata*
Caladenia macrostylis Fitzg.
(E) *Caladenia magniclavata* (Nicholls)
Hopper et A.P. Brown
Caladenia marginata Lindley
Caladenia nana Endl. ssp *nana*
Caladenia nana Endl. ssp *minta* (Fitg)
Hopper et A.P. Brown
Caladenia nivalis Hopper et A.P. Brown
Caladenia paludosa Hopper et A.P. Brown
Caladenia patersonii R. Br.
Caladenia pectinata R.S. Rogers
Caladenia pectinata ssp. *pectinata*
Caladenia pendens Hopper ined.
(E) *Caladenia pholcoidea* Hopper et A.P. Brown
Caladenia plicata Fitzg.
Caladenia radiata Nicholls
Caladenia reptans Lindley
(E) *Caladenia rhomboidiformis* (Coleman)
Clements et Hopper
Caladenia sericea Lindley
Caladenia serotina Hopper et A.P. Brown
Caladenia speciosa Hopper ined.
Caladenia viridescens Hopper et A.P. Brown
(E) *Caladenia winfieldii* Hopper et A.P. Brown
Calochilus robertsonii Benth.
Corybas despectans D.L. Jones et R.C. Nash
Corybas aff. *diemenicus* (Lindley) Rupp
Corybas recurva D. Jones
Corybas abditus D. Jones
Corybas aff. *unguiculatus* (R. Br.) H.G. Reichb.
Cryptostylis ovata R. Br.
Cyrtostylis huegelii Endl.
Cyrtostylis robusta D. Jones et M. Clements
Cyrtostylis tenuissima (Nicholls et Goadby)
D. Jones et M. Clements
Cyanicula deformis (R. Br) Hopper et A.P. Brown
Cyanicula gemmata (Lindley) Hopper et A.P. Brown
Cyanicula gertrudeae (Ostrey) Hopper et A.P. Brown
Cyanicula sericea (Lindley) Hopper et A.P. Brown
Diuris aff *amplissima* D. Jones
Diuris carinata Lindley
Diuris aff. *corymbosa* Lindley
Diuris drummondii Lindley
Diuris emarginata R. Br. var. *emarginata*
(E) *Diuris emarginata* var. *pauciflora* (R. Br.) A.S. George
Diuris filifolia Lindley
Diuris herbelii sp.
Diuris laevis Fitzg.
Diuris laxiflora Lindley
Diuris longifolia R. Br.
Diuris pauciflora R. Br.
Diuris setacea R. Br.
Drakaea elastica Lindley
Drakaea glyptodon Fitzg.
Drakaea livida Drumm.

Drakaea micrantha Hopper et A.P. Brown
Drakaea thynniphila A.S. George
Elythranthera brunonis (Endl.) A.S. George
Elythranthera emarginata (Lindley) A.S. George
Epiblema grandiflorum R. Br.
Eriochilus dilatatus Lindley ssp. *dilatatus*
Eriochilus dilatatus ssp. *multiflora* (Lindley) Hopper et A.P. Brown
Eriochilus scaber Lindley ssp. *scaber*
Eriochilus helenomos Hopper et A.P. Brown
Eriochilus tenuis Lindley
Eriochilus pulchellus Hopper et A.P. Brown
Eriochilus scaber ssp. *orbifolia* Hopper et A.P. Brown
Eriochilus valens Hopper et A.P. Brown
Gastrodia lacista sp.
Leporella fimbriata (Lindley) A.S. George
Leptoceras menziesii (R. Br.) Lindley
Rostranthus forestii (F. Muell) Hopper et A.P. Brown
Rostranthus nigricans (R. Br) Hopper et A.P. Brown
Lyperanthus serratus Lindley
Microtis alba R. Br.
Microtis atrata Lindley
(E) *Microtis globula* R. Bates
Microtis brownii H.G. Reibold
Microtis media R. Br. ssp. *media*
Microtis media ssp. *densiflora* (Benth) R. Bates
Microtis orbicularis R.S. Rogers
(E) *Microtis pulchella* R. Br.
Microtis rara R. Br.
Microtis unifolia (G. Forster) H.G. Reichb.
Microtis media ssp. *quadrata* R. Bates
* *Monadenia bracteata* (Sw.) T. Durand et Schinz
Paracaleana nigrita (Lindley) Blaxell
Paracaleana ultima Hopper et A.P. Brown
Praecoxanthus aphyllus (Benth) Hopper et A.P. Brown
Prasophyllum attenatum M. Clements
Prasophyllum brownii H.G. Reichb.
Prasophyllum calcicola R. Bates
Prasophyllum cyphochilum Benth.
Prasophyllum drummondii H.G. Reichb.
Prasophyllum elatum R. Br.
Prasophyllum fimbria H.G. Reichb.
Prasophyllum gibbosum R. Br.
(E) *Prasophyllum* aff. *gibbosum*
Prasophyllum giganteum Lindley
Prasophyllum hians H.G. Reichb.
Prasophyllum macrostachyum R. Br. var. *macrostachyum*
Prasophyllum aff. *macrostachyum* R. Br.
Prasophyllum odoratum R. Rogers
Prasophyllum ovale Lindley var. *triglochin* H.G. Reichb.
Prasophyllum parvifolium Lindley
Prasophyllum aff. *parvifolium* Lindley
Prasophyllum patens R. Br.
Prasophyllum sphumaeforme Fitg.
Prasophyllum regium R.S. Rogers
Prasophyllum triangulare Fitzg.
Pterostylis aspera D. Jones et M. Clements
Pterostylis barbata Lindley
Pterostylis aff. *dilatata* A.S. George
Pterostylis aff. *nana* R. Br.
Pterostylis aff. *plumosa* L. Cady
Pterostylis pyramidalis Lindley
Pterostylis recurva Benth.
Pterostylis robusta R. Rogers
Pterostylis rogersii E. Coleman
Pterostylis rogersii E. Coleman
(?E) *Pterostylis turfosa* Lindley
Pterostylis vittata Lindley
Pterostylis aff *vittata*
Thelymitra antennifera (Lindley) J.D. Hook.
Thelymitra aristata Lindley
Thelymitra campanulata
Thelymitra canaliculata R. Br.
Thelymitra cornicina H.G. Reichb.
Thelymitra crinita Lindley
Thelymitra cucullata Rupp.
Thelymitra flexuosa Endl.
Thelymitra fuscolutea R. Br. var. *fuscolutea*
Thelymitra aff. *fuscolutea* R. Br.
Thelymitra aff. *holmesii* Nicholls
Thelymitra aff. *jacksonii* Hopper et A.P. Brownii
Thelymitra longifolia J. Forst. et G. Forst.
Thelymitra maculata Hopper et A.P. Brown
Thelymitra mucida Fitzg.
Thelymitra aff *nuda* R. Br.
Thelymitra pauciflora R. Br.
Thelymitra aff. *pauciflora*
Thelymitra spiralis (Lindley) F. Muell.
Thelymitra stellata Lindley
Thelymitra tigrina R. Br.
Thelymitra variegata (Lindley) F. Muell.
Thelymitra villosa Lindley

DICOTYLEDONS

CASUARINACEAE
Allocasuarina decussata (Benth.) L. Johnson
Allocasuarina fraseriana (Miq.) L. Johnson
Allocasuarina huegeliana (Miq.) L. Johnson
Allocasuarina humilis (Otto et Dietr.) L. Johnson
Allocasuarina lehmanniana (Miq.) L. Johnson
Allocasuarina trichodon (Miq.) L. Johnson
Allocasuarina thuyoides (Miq.) L. Johnson

URTICACEAE
Parietaria debilis G. Forster
* *Soleirolia soleirolii* (Req.) Dandy
* *Urtica urens* L.

PROTEACEAE
Adenanthos apiculatus R. Br.
Adenanthos barbigerus Lindley
Adenanthos cuneatus Labill.
Adenanthos cunninghamii Meissner
(E) *Adenanthos detmoldii* F. Muell.
Adenanthos meisneri Lehm.
Adenanthos obovatus Labill.
Adenanthos sericeus Labill. ssp. *sericeus*
Banksia attenuata R. Br.
Banksia gardneri A.S. George
Banksia goodii R. Br.

Banksia grandis Willd.
Banksia ilicifolia R. Br.
Banksia littoralis R. Br.
Banksia meisneri Lehm. var. *adscendens* A.S. George
(E) *Banksia occidentalis* R. Br. ssp. *formosa* Hopper
Banksia R. Br. ssp. *occidentalis*
Banksia praemorsa Andrews
Banksia quercifolia R. Br.
Banksia seminuda (A.S. George) B.L. Rye
ssp. *seminuda*
(E) *Banksia seminuda* ssp. *remanens* Hopper
Banksia sphaerocarpa R. Br.
Banksia verticillata R. Br.
Conospermum acerosum Lindley
Conospermum caeruleum R. Br.
Conospermum capitatum R. Br.
Conospermum flexuosum R. Br.
Conospermum floribundum Benth.
Conospermum petiolare R. Br.
Conospermum teretifolium R. Br.
Conospermum triplinervium R. Br.
(E) *Conospermum* sp.
Dryandra armata R. Br.
Dryandra baxteri R. Br.
Dryandra bipinnatifida R. Br.
Dryandra cuneata R. Br.
Dryandra formosa R. Br.
Dryandra mimica A.S. George
Dryandra mucronulata R. Br.
Dryandra nivea (Labill.) R. Br.
Dryandra serra R. Br.
Dryandra sessilis (Knight) Domin
Dryandra sessilis (Knight) Domin ssp. nov.
Dryandra subpinnatifida C. Gardner
Franklandia fucifolia R. Br.
Franklandia triaristata Benth.
Grevillea bipinnatifida R. Br.
(E) *Grevillea brachystylis* Meissner
var. *australis* Keighery
Grevillea brownii Meissner
Grevillea bronwenae Keighery
Grevillea cirsiifolia Meissner
Grevillea diversifolia Meissner ssp. *subteriser icata*
Grevillea drummondii Meissner
ssp. *centristigma* Macgillivray
Grevillea fasciculata R. Br.
Grevillea fistulosa A.S. George
ssp. nov.
Grevillea fuscolutea Keighery
Grevillea leptobotrya Meissner
Grevillea manglesioides Meissner
ssp. *manglesioides*
Grevillea manglesioides
ssp. *papillosa* McGillivray
Grevillea occidentalis R. Br.
Grevillea pilulifera (Lindley) Druce
Grevillea pulchella (R. Br.) Meissner
Grevillea quercifolia R. Br.
Grevillea sp. *ripicola* A.S. George
Grevillea trifida (R. Br.) Meissner
Hakea amplexicaulis R. Br.
Hakea ceratophylla (Smith) R. Br.
Hakea cyclocarpa Lindley
Hakea elliptica (Smith) R. Br.
Hakea falcata R. Br.
Hakea florida R. Br.
Hakea lasiantha R. Br.
Hakea lasianthoides B.L. Rye
Hakea linearis R. Br.
Hakea lissocarpha R. Br.
Hakea oleifolia (Smith) R. Br.
Hakea prostrata R. Br.
Hakea ruscifolia Labill.
Hakea suaveolens R. Br.
Hakea sulcata R. Br.
Hakea trifurcata (Sm.) R. Br.
Hakea undulata R. Br.
Hakea varia R. Br.
Isopogon attenuatus R. Br.
var *attenuatus*
Isopogon axillaris R. Br.
Isopogon buxifolius R. Br.
Isopogon formosus R. Br.
Isopogon uncinatus R. Br.
Isopogon sphaerocephalus Lindley
Isopogon teretifolius R. Br.
Lambertia inermis R. Br.
Lambertia multiflora Lindley
Lambertia orbifolia C. Gardner
Lambertia propinqua R. Br.
Lambertia rariflora Meissner
Lambertia uniflora R. Br.
Persoonia elliptica R. Br.
Persoonia graminea R. Br.
Persoonia longifolia R. Br.
Persoonia microcarpa R. Br.
Persoonia saccata R. Br.
Persoonia teretifolia
Petrophile acicularis R. Br.
Petrophile divaricata R. Br.
Petrophile diversifolia R. Br.
Petrophile linearis R. Br.
Petrophile longifolia R. Br.
Petrophile media R. Br.
Petrophile rigida R. Br.
Petrophile serruriae R. Br.
Petrophile squamata R. Br.
ssp. "A" (short leaved form)
Petrophile squamata R. Br.
ssp. "B" (fine leaved form)
Stirlingia latifolia (R. Br.) Steudel
Stirlingia simplex Lindley
Stirlingia tenuifolia (R. Br.) Steudel
Strangea stenocarpoides (F. Muell. ex Benth.)
C. Gardner
Synaphea favosa R. Br.
Synaphea gracillima Lindley
Synaphea petiolaris R. Br.
Synaphea polymorpha R. Br.
Synaphea preissii Meissner
Synaphea reticulata (Smith) C. Gardner
Xylomelum occidentale R. Br.

SANTALACEAE
(E) *Choretrum lateriflorum* R. Br.
Exocarpus odoratus (Miq.) A.DC.
Exocarpus sparteus R. Br.
Leptomeria cunninghamii Miq.
Leptomeria lehmaniannii Miq.
Leptomeria ericoides Miq.
Leptomeria pauciflora R. Br.
Leptomeria scrobiculata R. Br.
Leptomeria squarulosa R. Br.
Leptomeria spinosa (Miq.) A.DC.
Santalum spicatum (R. Br.) A.DC.

OLACACEAE
Olax phyllanthi (Labill.) R. Br.
Olax benthamiana Miq.

LORANTHACEAE
Amyema miquelii (Lehm. ex Miq.) Tieghem
Nuytsia floribunda (Labill.) R. Br.

POLYGONACEAE
* *Emex australis* Steinh.
* *Fagopyrum esculentum* Moench
Muehlenbeckia adpressa (Labill.) Meissner
* *Polygonum attenuatum* R. Br.
Polygonum hydropiper L.
Polygonum prostratum R. Br.
Polygonum salicifolium Brouss. ex Willd.
* *Rumex acetosella* L.
Rumex brownii Campdera
* *Rumex conglomeratus* Murr.
* *Rumex crispus* L.
* *Rumex frutescens* Thouars
* *Rumex pulcher* L. ssp. *pulcher*

CHENOPODIACEAE
Atriplex bunburyana F. Muell.
Atriplex cinerea Poiret
* *Atriplex hortensis* L.
Atriplex hypoleuca Nees
Atriplex isatidea Moq.
Atriplex paludosa R. Br.
ssp. *baudinii* (Moq.) Aellen
* *Atriplex prostrata* M. Boucher ex DC.
* *Chenopodium album* L.
* *Chenopodium ambrosioides* L.
* *Chenopodium glaucum* L.
* *Chenopodium murale* L.
Chenopodium pumilio R. Br.
Halosarcia halocnemoides (Nees) P.G. Wilson
ssp. *halocnemoides*
Halosarcia indica (Willd.) P.G. Wilson
ssp. *bidens* (Nees) P.G. Wilson
Halosarcia pergranulata (J.M. Black) P.G. Wilson
ssp. *pergranulata*
Halosarcia pterygosperma (Black) P.G. Wilson
ssp. *pterygosperma*
Halosarcia syncarpa P.G. Wilson
Maireana brevifolia (R. Br.) P.G. Wilson
Maireana oppositifolia (F. Muell.) P.G. Wilson
Rhagodia baccata (Labill.) Moq. ssp. *baccata*
Rhagodia baccata (Labill) Moq. ssp. *choica*
Rhagodia crassifolia R. Br.
Salsola kali L. ssp. *kali*
Sarcocornia blackiana (Ulbr.) A.J. Scott
Sarcocornia quinqueflora (Bunge ex Ung-Sternb.)
A.J. Scott ssp. *quinqueflora*
Suaeda australis (R. Br.) Moq.
Threlkeldia diffusa R. Br.

AMARANTHACEAE
Alternanthera nodiflora R. Br.
* *Amaranthus albus* L.
Hemichroa *diandra* R. Br.
Ptilotus sericostachyus (Nees) F. Muell.
Ptilotus declinatus Nees
Ptilotus stirlingii var. *laxus* (Benth.) Benl.
Ptilotus stirlingii (Lindley) F. Muell. var. *stirlingii*

GYROSTEMONACEAE
Gyrostemon sheathii W. Fitzg.
Gyrostemon thesioides (J.D. Hook.) A.S. George

PHYTOLACCACEAE
* *Phytolacca octandra* L.

AIZOACEAE
* *Carpobrotus aequilaterus* (Haw.) N.E. Br.
* *Carpobrotus edulis* (L.) L. Bolus
Carpobrotus virescens (Haw.) Schwantes
* *Drosanthemum candens* (Haw.) Schwantes
* *Lampranthus glaucus* (L.) N.E. Br.
* *Mesembryanthemum crystallinum* L.
* *Tetragonia decumbens* Miller
Tetragonia implexicoma (Miq.) J.D. Hook.
Tetragonia tetragonoides (Pallas) Kuntze

PORTULACACEAE
Calandrinia brevipedata F. Muell.
Calandrinia calyptrata J.D. Hook.
Calandrinia corrigioloides F. Muell. ex Benth.
Calandrinia granulifera Benth.
Calandrinia liniflora Fenzl

CARYOPHYLLACEAE
* *Cerastium glomeratum* Thuill.
* *Cerastium semidecandrum* L.
* *Corrigiola litoralis* L.
* *Lychnis coronaria* L.
* *Moenchia erecta* (L.) P. Gaertner
* *Petrorhagia velutina* (Guss.) P. Ball et Heyw.
* *Polycarpon tetraphyllum* (L.) L.
* *Sagina maritima* G. Don
* *Sagina procumbens* L.
* *Silene gallica* L. var. *gallica*
Silene gallica L. var. *quinquevulnera*
(L) Mert et Koch
* *Silene nocturna* L.
* *Spergula diandra* Heldr. et Sart
* *Stellaria media* (L.) Vill.
* *Vaccaria pyramidata* L.

RANUNCULACEAE

Clematis microphylla DC.
Clematis pubescens Huegel ex Endl.
Ranunculus colonorum Endl.
* *Ranunculus muricatus* L.
Ranunculus rivularis Banks et Sol. ex DC.

LAURACEAE

Cassytha glabella R. Br. forma *casurinae* (Nees) J.Z. Weber
Cassytha flava nees.
Cassytha melantha R. Br.
Cassytha micrantha Meissner
Cassytha pomiformis Nees
Cassytha racemosa Nees forma *racemosa*
Cassytha racemosa forma *pilosa* (Benth.) J.Z. Weber

FUMARIACEAE

* *Fumaria capreolata* L.
* *Fumaria muralis* Sonder ex Koch
* *Fumaria officinalis* L.

BRASSICACEAE

* *Brassica rapa* L. ssp. *sylvestris* (L.) Janden
* *Brassica tournefortii* Gouan
* *Cakile maritima* Scop.
* *Capsella bursa-pastoris* (L.) Medikus
* *Cardamine hirsuta* L.
Cardamine paucijuga Turcz.
* *Coronopus didymus* (L.) Smith
* *Diplotaxis muralis* (L.) DC.
* *Heliophila pusilla* L.f.
* *Hirschfieldia incana* (L.) Lag-Foss.
* *Hymenolobus procumbens* (L.) Nutt.
* *Lepidium africanum* (Burm. f.) DC.
Lepidium foliosum Desv.
Lepidium pseudohysopfolium Hewson
Lepidium rotundum (Desv.) DC.
* *Nasturtium officinale* R. Br.
* *Rapistrum rugosum* (L.) All.
* *Raphanus raphanistrum* L.
Rorippa dictyosperma (F. Muell.) L. Johnson
* *Sinapis arvensis* L.
* *Sisymbrium irio* L.
* *Sisymbrium orientale* L.
Stenopetalum robustum Endl.

RESEDACEAE

* *Reseda luteola* L.

DROSERACEAE

Drosera bulbosa Hook.
Drosera dichrosepala Turcz.
Drosera erythrorhiza Lindley ssp. *erythrorhiza*
Drosera erythrorhiza Lindley ssp. *squarrosa*
Drosera gigantea Lindley
Drosera glanduligera
Drosera hamiltonii C.R.P. Andrews
Drosera huegelii Endl.
Drosera macrantha Endl. ssp. *macrantha*
Drosera menziesii R. Br. ssp. *menziesii*
Drosera microphylla Endl.
Drosera modesta Diels
Drosera myriantha Planchon
Drosera neesii Lehm. ssp. *neesii*
Drosera occidentalis Morrison ssp. *australis*
Drosera omissa Diels
Drosera pallida Lindley
Drosera platypoda Turcz.
Drosera platystigma Lehm.
Drosera pulchella Lehm.
Drosera pygmaea DC.
Drosera ramellosa Lehm.
Drosera stolonifera Endl. ssp. *stolonifera*
Drosera subhirtella Planchon

CRASSULACEAE

* *Aoenium castello-pavoniae* Bolle
* *Cotyledon orbiculare* L.
Crassula colorata (Nees) Ostenf. var. *colorata*
Crassula colorata var. *acuminata* (Reader) Tolken
Crassula decumbens Thunb. var. *decumbens*
* *Crassula glomerata* P. Bergius
* *Crassula natans* Thunb. var. *minus* (Eckl. et Zeyh.) Rowley
Crassula sieberiana (J.A. and J.H. Schultes) Druce ssp. *tetramera* Tolken
* *Crassula thunbergiana* J.A. Schultes ssp. *thunbergiana*
* *Crassula tetragona* L. ssp. *robusta* (Toelken) Toelken

CEPHALOTACEAE

Cephalotus follicularis Labill.

SAXIFRAGACEAE

Eremosyne pectinata Endl.

PITTOSPORACEAE

Billardiera candida (Huegel ex Endl.) E.M. Bennett
Billardiera coerulea-punctata
Billardiera drummondiana (Putterl.) E.M. Bennett var. *drummondiana*
Billardiera erubescens (Putterl.) E.M. Bennett
Billardiera floribunda (Putterl.) F. Muell.
Billardiera laxiflora (Benth.) E.M. Bennett
Billardiera parviflora DC. var. *parviflora*
Billardiera sericea (Turcz.) E.M. Bennett
Billardiera variifolia DC.
Cheiranthera preissiana Putterl. var. *planifolia* E.M. Bennett
(E) *Sollya drummondii* Morren
Sollya heterophylla Lindley

ROSACEAE

* *Acaena echinata* Nees var. *retrosumpilosa* (Bitter) Orch.
* *Acaena novae-zelandiae* Kirk
* *Cotoneaster glaucophyllus* L.
* Rosa *chinensis* Jacq. x Rosa *multiflora* Thunb. ex Murray
* *Rosa rubiginosa* L.
* *Rubus discolor* Weihe et Nees
* *Rubus selmeri* Lindeb. ex F. Aresch.

* *Rubus ulmifolius* Schott
* *Sanguisorba minor* Scop.

MIMOSACEAE
Acacia acuminata Benth.
Acacia acutifolia Maiden et Blakely
Acacia alata R. Br.
Acacia assimilis S. Moore
Acacia ataxiphylla Benth.
Acacia baxteri Benth.
Acacia biflora R. Br.
Acacia browniana Wendl. var. *browniana*
Acacia browniana Wendl. var. *obscura* (A.DC.) B.R. Maslin
Acacia browniana Wendl. var. *endlicheri*
Acacia cochlearis (Labill.) H. Wendl.
Acacia costata Benth.
Acacia crassiuscula Wendl.
Acacia crispula
Acacia cyclops Cunn. ex Don.
* *Acacia decurrens* (Wendl.) Willd.
Acacia divergens Benth.
* *Acacia dealbata* Link
Acacia drummondii Lindley ssp. *elegans* B.R. Maslin
Acacia extensa Lindley
Acacia fragilis Maiden et Blakely
Acacia gilbertii Meissner
Acacia hastulata Smith in Rees
Acacia huegelii
Acacia incurva Benth.
Acacia inops Rareflora
Acacia insolita E. Pritzel
Acacia laricina Meissner
Acacia latipes Benth.
Acacia leioderma Maslin
Acacia lateriticola Maslin
Acacia littorea Maslin
Acacia luteola Maslin
Acacia microbotrya Benth.
Acacia mooreana W. Fitzg.
Acacia multispicata Benth.
* *Acacia melanoxylon* R. Br.
Acacia myrtifolia (Smith) Willd.
Acacia nervosa DC.
Acacia obovata Benth.
Acacia paradoxa DC.
(E) *Acacia pentadenia* Lindley
Acacia aff. pentadenia (Annels 3700)
Acacia preissiana (Meissner) Maslin
Acacia prismifolia E. Pritzel
Acacia pulchella R. Br. var. *pulchella*
Acacia pulchella R. Br. var. *glaberrima* Meissner
Acacia pulchella R. Br. var. *goadbyi* (Domin.) B.R. Maslin
* *Acacia pycnantha* Benth.
Acacia pycnocephala Maslin
Acacia robinae Maslin
Acacia rostellifera Benth.
Acacia saligna (Labill.) H.L. Wendl.
(E) *Acacia scalpelliformis* Meissner
Acacia semitrullata Maslin
Acacia stenoptera Benth.
Acacia subcaerulea Lindley
(E) *Acacia subracemosa* Maslin
Acacia sulcata R. Br.
(E) *Acacia tayloriana* F. Muell.
Acacia triptycha F. Muell. ex Benth.
Acacia tetragonocarpa Meissner
Acacia uliginosa Maslin
Acacia urophylla Benth. ex Lindley
Acacia varia Maslin var. *varia*
Acacia willdenowiana H.L. Wendl.
Paraserianthes lophantha (Willd.) I. Nielsen

CAESALPINIACEAE
Labichea punctata Benth.

PAPILIONACEAE
Aotus carinata Meissner
Aotus sp.
Aotus intermedia Meissner
(E) *Aotus passerinoides* Meissner
Aotus genistoides Turcz.
Aotus gracillima Meissner
Aotus procumbens Meissner
Aotus tietkinsii F. Muell.
Bossiaea aquifolium Benth.
Bossiaea dentata (R. Br.) Benth.
(E) *Bossiaea disticha* Lindley
Bossiaea eriocarpa Benth.
Bossiaea laidlawiana Tovey et P. Morris
Bossiaea linophylla R. Br.
Bossiaea ornata (Lindley) Benth.
Bossiaea rufa R. Br.
Bossiaea webbii F. Muell.
Brachysema praemorsum Meissner
Brachysema sericeum (Smith) Domin.
Burtonia conferta DC.
Burtonia scabra (Smith) R. Br.
Burtonia villosa Meissner
Chorizema aciculare (DC.) C. Gardner
Chorizema cordatum Lindley
Chorizema ilicifolium Labill.
Chorizema diversifolium DC.
Chorizema glycinifolium (Smith) Druce
Chorizema reticulatum Meissner
Chorizema rhombeum R. Br.
(E) *Chorizema* sp (varium)
* *Cytisus proliferus* L.f.
Daviesia alternifolia Endl.
Daviesia angulata Benth.
Daviesia benthamii Meissner
Daviesia brachyphylla M.D. Crisp
Daviesia colleloides Meissner
Daviesia cordata Smith
Daviesia decurrens Meissner
Daviesia divaricata Benth.
Daviesia flexuosa Benth.
Daviesia horrida Preiss ex Meissner
Daviesia gracilis M.D. Crisp
Daviesia incrassata Smith
Daviesia inflata M.D. Crisp
Daviesia longifolia Benth.

Daviesia oppositifolia Endl.
Daviesia polyphylla Benth.
Daviesia preissii Meissner
Daviesia rhombifolia Meissner
Dillwynia cinerascens R. Br. ex Sims
Dillwynia uncinata (Turcz.) J. Black
* *Dipogon lignosus* (L.) Verdc.
Euchilopsis linearis (Benth.) F. Muell.
Eutaxia densifolia Turcz.
Eutaxia epacridoides Meissner
Eutaxia obovata (Labill.) C. Gardner
Eutaxia parvifolia Benth. Endl.
Eutaxia virgata Benth.
Gastrolobium bilobum R. Br.
Gastrolobium brownii Meissner
Gastrolobium callistachys Meissner
Gastrolobium spinosum Benth.
Gastrolobium forrestii Ewart
Gastrolobium villosum Benth.
* *Genista canariensis* L.
* *Genista linifolia* L.
Gompholobium amplexicaule Meissner
Gompholobium aristatum Benth.
Gompholobium burtonioides Meissner
Gompholobium capitatum Cunn.
Gompholobium knightianum Lindley
Gompholobium marginatum R. Br.
Gompholobium ovatum Meissner
Gompholobium polymorphum R. Br.
Gompholobium tomentosum Labill.
Gompholobium venustum R. Br.
Goodia lotifolia Salisb.
Hardenbergia comptoniana (Andrews) Benth.
Hovea chorizemifolia (Sweet) DC.
Hovea stricta Meissner
Hovea trisperma Benth. var.
Hovea elliptica (Smith) DC.
Isotropis cuneifolia (Smith) Benth. ex B.D. Jackson
Jacksonia alata Benth.
Jacksonia aphylla (Turcz.) Druce
Jacksonia furcellata (Bonpl.) DC.
Jacksonia horrida DC.
Jacksonia mollissima W. Fitzg.
Jacksonia spinosa (Labill.) R. Br.
Jacksonia sternbergiana Huegel
(E) *Jansonia formosa* Kipp. ex Lindley
Kennedia carinata (Benth.) Domin
Kennedia coccinea Vent.
(E) *Kennedia glabrata* (Benth.) Lindley
(E) *Kennedia macrophylla* (Meissner) Benth.
Kennedia microphylla Meissner
Kennedia prostrata R. Br. .
Kennedia stirlingii Lindley
Latrobaea brunonis (Benth.) Meissner
Latrobea diosmifolia Benth.
Latrobea genistoides (Meissner)
Latrobea hirtella (Turcz.) Benth.
Latrobea tenella (Meissner) Benth. var. *tenella*
* *Lathyrus tingitanus* L.
* *Lathyrus sylvestris* L.
* *Lotus angustissimus* L.
* *Lotus suaveolens* Pers.
* *Lotus uliginosus* Schk.
* *Lupinus luteus* L.
* *Medicago arabica* (L.) Hudson
* *Medicago lupulina* L.
* *Medicago polymorpha* L. var. *brevispina* (Benth.) Heyn.
* *Melilotus indica* (L.) All.
Mirbelia dilatata R. Br.
Mirbelia ovata Meissner
Mirbelia spinosa Benth.
Nemeia hookeri (Meissner) M.D. Crisp
Nemeia spathulata
* *Ornithopus compressus* L.
* *Ornithopus pinnatus* (Miller) Druce
Oxylobium carinatum (Meissner) Benth.
Oxylobium coriaceum (Smith) C. Gardner
Oxylobium drummondii Meissner
Oxylobium lanceolatum (Vent.) Druce
Oxylobium linearifolium (G. Don.) Domin.
Oxylobium spathulatum (Meissner) Benth.
Phyllota barbata Benth.
* *Psoralea pinnata* L.
Pultenaea adunca Turcz.
Pultenaea aspalanthoides Meissner
Pultenaea barbata C.R.P. Andrews
Pultenaea calyeina (Turcz) Benth.
Pultenaea drummondii Meissner
Pultenaea ericifolia Benth.
Pultenaea ochreata Meissner
Pultenaea pinifolia Meissner
Pultenaea reticulata (Smith) Benth.
Pultenaea skinneri F. Muell.
Pultenaea strobilifera Meissner
Pultenaea verruculosa Turcz.
Pultenaea vestita R. Br.
Sphaerolobium alatum Benth.
Sphaerolobium fornicatum Benth.
Sphaerolobium grandiflorum (R. Br.) Benth.
Sphaerolobium linophyllum (Huegel) Benth.
Sphaerolobium macranthum Meissner
Sphaerolobium medium R. Br.
Sphaerolobium nudiflorum (Meissner) Benth.
Sphaerolobium racemulosum Benth.
Sphaerolobium scabriusculum Meissner
Sphaerolobium vimineum Smith
Templetonia retusa (Vent.) R. Br.
* *Trifolium arvense* L.
* *Trifolium campestre* Schreber
* *Trifolium cernuum* Brot.
* *Trifolium dubium* Sibth.
* *Trifolium glomeratum* L.
* *Trifolium hirtum* All.
* *Trifolium ligusticum* Balbis ex Lois.
* *Trifolium repens* L.
* *Trifolium striatum* L.
* *Trifolium tomentosum* L.
* *Ulex europaeus* L.
* *Vicia hirsuta* (L.) Gray
* *Vicia sativa* L. ssp. *sativa*
* *Vicia sativa* ssp. *nigra* (L.) Ehrh.
Viminaria juncea (Schrader et Wendl.) Hoffsgg.

GERANIACEAE

* *Erodium botrys* (Cav.) Bertol
* *Erodium cicutarium* (L.) L'Her.
Erodium cygnorum Nees ssp. *cygnorum*
* *Geranium dissectum* L.
Geranium drummondii Carolin
Geranium retrorsum L'Her. ex DC.
Geranium solanderi Carolin
Pelargonium alchemilloides (L) L Her
ssp. *alchemilloides*
(Not in Green)
Pelargonium australe Willd.
* *Pelargonium capitatum* (L.) L.Her.
Pelargonium drummondii Turcz.
Pelargonium littorale Huegel

OXALIDACEAE

* *Oxalis corniculata* L.
Oxalis perennans Haw.
* *Oxalis flava* L.
* *Oxalis incarnata* L.
* *Oxalis pes-caprae* L.
* *Oxalis polyphylla* Jacq.
* *Oxalis purpurea* L.

LINACEAE

Linum marginale Cunn. ex Planchon
* *Linum trigynum* L.

ZYGOPHYLLACEAE

Nitraria billardierei DC.

RUTACEAE

Asterolasia pallida Benth.
Asterolasia squamuligera (Hook.) Benth.
Boronia alata Smith
Boronia albiflora R. Br. ex Benth.
Boronia crenulata Sm. var. *crenulata*
Boronia crenulata var. *pubescens* Benth.
Boronia crassipes Bartling
Boronia defoliata F. Muell.
Boronia denticulata Smith
Boronia dichotoma Lindley
Boronia fastigiata Bartling
Boronia gracilipes F. Muell.
Boronia heterophylla F. Muell.
Boronia juncea Bartling
Boronia megastigma Nees ex Bartling
Boronia molloyae J. Drumm.
Boronia nematophylla F. Muell.
Boronia pulchella Turcz.
Boronia sp. aff. *juncea*
Boronia spathulata Lindley
Boronia sp. aff. *spathulata*
Boronia stricta Bartling
Boronia subsessilis Benth.
Boronia tenuis (Lindley) Benth.
Boronia virgata P.G. Wilson
Chorilaena quercifolia Endl.
Crowea angustifolia Smith var. *angustifolia*
Crowea angustifolia var. *dentata*
(Benth.) P.G. Wilson
Diplolaena dampieri Desf.
Diplolaena drummondii (Benth.) Ostenf.
Diplolaena microcephala Bartling
var. *microcephala*
Eriostemon nodiflorus Lindley var. *nodiflorus*
Eriostemon spicatus A. Rich.
Phebalium anceps DC.
Phebalium rude Bartling ssp. *rude*

TREMANDRACEAE

Platytheca galioides Steetz
Tetratheca affinis Endl.
(E) *Tetratheca elliptica* J. Thompson
(E) *Tetratheca filiformis* Benth.
Tetratheca hirsuta Lindley
Tetratheca hispidissima Steetz
Tetratheca setigera Endl.
Tremandra diffusa R. Br. ex DC.
Tremandra stelligera R. Br.

POLYGALACEAE

Comesperma calymega Labill.
Comesperma ciliatum Steetz
Comesperma confertum Labill.
Comesperma flavum DC.
Comesperma nudiusculum DC.
Comesperma virgatum Labill.
Comesperma volubile Labill.
* *Polygala myrtifolia* L.
* *Polygala virgata* Thunb.

EUPHORBIACEAE

Adriana quadripartita (Labill.) Gaudich
Amperea ericoides Adr. Juss.
Amperea micrantha Benth.
Amperea protensa Nees.
Amperea volubilis F. Muell. ex Benth.
Amperea sp. (CJR 227)
Beyeria viscosa (Labill.) Miq.
* *Euphorbia peplus* L.
* *Euphorbia paralias* L.
* *Euphorbia helioscopia* L.
* *Mercurialis annua* L.
Monotaxis grandiflora Endl.
Monotaxis occidentalis Endl.
Poranthera huegelii Klotzsch
Poranthera microphylla Brongn.
Phyllanthus calycinus Labill.
Phyllanthus aff. *scaber* Klotzsch)
Ricinocarpus glaucus Endl.
* *Ricinus communis* L.

CALLITRICHACEAE

* *Callitriche stagnalis* Scop.

STACKHOUSIACEAE

Stackhousia pubescens A. Rich.
Tripterococcus brunonis Endl.
(E) *Tripterococcus* sp. nov.

SAPINDACEAE

Dodonaea aptera Miq.

Dodonaea ceratocarpa Endl.
Dodonaea trifida F. Muell.

RHAMNACEAE
Cryptandra arbutiflora Fenzl
Cryptandra pungens Steudel
Cryptandra tubulosa Fenzl
Pomaderris myrtilloides Fenzl
Spyridium globulosum (Labill.) Benth.
(E) *Spyridium spadiceum* (Fenzl) Benth.
Trymalium floribundum Steudel
Trymalium ledifolium Fenzl var. *ledifolium*
(E) *Trymalium* aff. *floribundum* (R.D. Royce 4286)

MALVACEAE
Guichenotia ledifolia Gay.
* *Lavatera arborea* L.
Lavatera plebeia Sims var. *tomentosa* Hook. f.
Lavatera plebeia var. *pleibeia*
* *Malva parviflora* L.
* *Modiola caroliniana* (L.) Don.
Sida hookeriana Miq.

STERCULIACEAE
Guichenotia ledifolia Gay
Lasiopetalum cordifolium Endl.
(E) *Lasiopetalum floribundum* Benth. ssp. nov.
Rulingia corylifolia R.A. Graham
Rulingia cygnorum (Steudel)
C. Gardner var. *cygnorum*
Rulingia grandiflora Endl.
Rulingia parviflora Endl.
Thomasia cognata Steudel
Thomasia brachystachys Turez.
Thomasia discolor Steudel
Thomasia foliosa Gay
Thomasia grandiflora Lindley
Thomasia laxiflora Benth.
Thomasia macrocalyx Steudel
Thomasia pauciflora Lindley
Thomasia multiflora E. Pritzel
Thomasia purpurea (Aiton) Gay
(E) *Thomasia quercifolia* (Andrews) Gay
Thomasia rhynchocarpa Turcz.
(E) *Thomasia solanacea* Gay
Thomasia triloba Turcz.
Thomasia triphylla (Labill.) Gay

DILLENIACEAE
Hibbertia acerosa (R. Br. ex DC.) Benth.
Hibbertia amplexicaulis Steudel
Hibbertia commutata Steudel
Hibbertia cuneiformis (Labill.) Smith
Hibbertia cunninghamii W.T. Aiton ex Hook.
Hibbertia furfuracea (R. Br. ex DC.) Benth.
Hibbertia grossulariifolia (Salisb.) Salisb.
Hibbertia glabberima F. Muell
Hibbertia glomerata Benth.
Hibbertia hypericoides (DC.) Benth.
Hibbertia inconspicua Ostenf.
Hibbertia lasiopus Benth
Hibbertia microphylla Steud.
Hibbertia pachyrrhiza Steudel
Hibbertia perfoliata Endl.
Hibbertia pulchra Ostenf.
Hibbertia quadricolor Domin
Hibbertia racemosa (Endl.) Gilg
Hibbertia rhadinopoda F. Muell.
Hibbertia serrata Hotchk.
Hibbertia silvestris Diels
Hibbertia stellaris Endl.
Hibbertia subvaginata (Benth) F. Muell.

CLUSIACEAE
Hypericum gramineum G. Forster
Hypericum japonicum Thunb.
* *Hypericum perforatum* L.

FRANKENIACEAE
Frankenia pauciflora DC.
Frankenia tetrapetala Labill.

VIOLACEAE
Hybanthus calycinus (DC. ex Ging.) F. Muell.
Hybanthus debilissimus F. Muell.
Hybanthus floribundus (Lindley) F. Muell.
ssp. *floribundus*
Hybanthus volubilis E.M. Bennett
* *Viola odorata* L.

THYMELEACEAE
Pimelea angustifolia R. Br.
Pimelea argentea R. Br.
Pimelea brevifolia R. Br. ssp. *brevifolia*
Pimelea ciliata B.L. Rye ssp. *ciliata*
Pimelea cracens B.L. Rye ssp. *cracens*
Pimelea cracens ssp. *glabra* Rye
Pimelea clavata Labill.
Pimelea ferruginea Labill.
Pimelea hispida R. Br.
Pimelea imbricata R. Br. ssp. *imbricata*
Pimelea imbricata ssp. *piligera* Rye
Pimelea lanata R. Br.
Pimelea lehmanniana Meissner ssp. *lehmanniana*
Pimelea longiflora R. Br. ssp. *longiflora*
Pimelea preissii Meissner
Pimelea rosea R. Br.
Pimelea spectabilis Lindley
Pimelea suaveolens Meissner ssp. *suaveolens*
Pimelea sylvestris R. Br.
Pimelea tinctoria Meissner

LYTHRACEAE
* *Lythrum hyssopifolia* L.

MYRTACEAE
Actinodium cunninghamii Schauer
Agonis flexuosa (Sprengel) Schauer
Agonis hypericifolia Schauer
Agonis juniperina Schauer
Agonis linearifolia (DC.) Schauer
Agonis marginata (Labill.) Schauer
Agonis parviceps Schauer
Astartea clavulata Turcz.

Astartea aff. *fascicularis* (Labill.) DC.
Astartea sp. nov. (G.K. 970)
Baeckea arbuscula R. Br. ex Benth.
Baeckea astarteoides Benth.
Baeckea blacketii F. Muell.
Baeckea camphorosmae Endl.
Baeckea pygmaea R. Br. ex Benth.
Beaufortia anisandra Schauer
Beaufortia decussata R. Br..
Beaufortia micrantha Schauer
Beaufortia sparsa R. Br.
Beaufortia squarrosa Schauer
Callistemon glaucus (Bonpl.) Sweet
Calothamnus gracilis R. Br.
Calothamnus graniticus ssp. *graniticus*
Calothamnus lateralis Lindley
Calothamnus preissii Schauer
Calothamnus sanguineus Labill.
Calothamnus schaueri Lehm.
Calytrix acutifolia (Lindley) Craven
Calytrix asperula (Schauer) Benth.
Calytrix birdii (F. Muell.) B.D. Jackson
Calytrix flavescens Cunn.
Calytrix leschenaultii (Schauer) Benth.
Calytrix tenuiramea (Turcz.) Benth.
Calytrix tetragona Labill.
Chamelaucium ciliatum Desf.
(E) *Chamelaucium* sp.
Darwinia citriodora (Endl.) Benth.
Darwinia diosmoides (DC.) Benth.
Darwinia forrestii F. Muell.
Darwinia oederoides (Turcz.) Benth.
Darwinia vestita (Endl.) Benth.
Eremaea pauciflora (Endl.) Druce
Eucalyptus angulosa Schauer
(E) *Eucalyptus brevistylis* Brooker
Eucalyptus calophylla Lindley
(E) *Eucalyptus calcicola* Brooker
Eucalyptus conferruminata D.J. Carr et S.G.M. Carr
Eucalyptus cornuta Labill.
Eucalyptus decipiens Endl.
Eucalyptus decurva F. Muell.
Eucalyptus diversicolor F. Muell.
Eucalyptus doratoxylon F. Muell.
Eucalyptus drummondii Benth.
Eucalyptus falcata Turcz.
(E) *Eucalyptus ficifolia* F. Muell.
Eucalyptus goniantha Turcz. ssp. nov.
(E) *Eucalyptus guilfoylei* Maiden
Eucalyptus haematoxylon Maiden
(E) *Eucalyptus jacksonii* Maiden
Eucalyptus loxophleba Benth.
Eucalyptus marginata Donn ex Smith
Eucalyptus misselis Brooker and Hopper in edit
Eucalyptus megacarpa F. Muell.
Eucalyptus occidentalis Endl.
Eucalyptus patens Benth.
Eucalyptus phylacia Hopper et A.P. Brown
Eucalyptus rudis ssp. nov.
Eucalyptus redunca Schauer
Eucalyptus rudis Endl.
Eucalyptus rudis ssp. nov.
Eucalyptus staeri (Maiden) Kessell et C.Gardner
Eucalyptus wandoo Blakely
Homalospermum firmum Schauer
Hypocalymma angustifolium Endl.
Hypocalymma cordifolium (Lehm.) Schauer
Hypocalymma ericifolium Benth.
Hypocalymma robustum Endl.
Hypocalymma strictum Schauer
(E) *Hypocalymma* sp.
Kunzea ericifolia (Smith) Heynh.
Kunzea aff. *micrantha* Schauer
Kunzea spicata S. Moore
Kunzea recurva Schauer var. *recurva*
Kunzea recurva Schauer var *melaleucoides*
(E) *Kunzea sulphurea* Tovey et P. Morris
Kunzea ericifolia (Smith) Heynh.
Leptospermum erubescens Schauer
* *Leptospermum laevigatum* (Gaertner) F. Muell.
Melaleuca acerosa Schauer
(?E) *Melaleuca basicephala* Benth.
Melaleuca baxteri Benth.
Melaleuca bracteosa Turcz.
Melaleuca cuticularis Labill.
Melaleuca densa R. Br.
Melaleuca diosmifolia Andrews
Melaleuca huegelii Endl.
Melaleuca incana R. Br.
Melaleuca lanceolata Otto
Melaleuca lateritia Otto. A. Dietr.
Melaleuca laxiflora Turcz.
Melaleuca leptoclada Benth
Melaleuca microphylla Smith
Melaleuca micromera Schauer
Melaleuca pauciflora Turcz.
Melaleuca pentagona Labill.
Melaleuca polygaloides Schauer
Melaleuca preissiana Schauer
Melaleuca rhaphiophylla Schauer
Melaleuca scabra R. Br. var *trichophylla*
Melaleuca seriata Lindley
Melaleuca spathulata Schauer
Melaleuca striata Labill.
Melaleuca thymoides Labill.
Melaleuca viminea Lindley
Melaleuca violacea Schauer
(E) *Pericalymma crassipes* (Endl.) Schauer
Pericalymma ellipticum (Endl.) Schauer
Scholtzia sp.
Thryptomene saxicola (Cunn. ex Hook.) Schauer
Thryptomene aff. *hyporhytis* Turcz.
Verticordia acerosa Lindley
Verticordia densiflora Lindley
Verticordia fimbrilepis Turcz. ssp. *australis*
Verticordia habrantha Schauer
Verticordia lehmanii Schauer
Verticordia lindleyi Schauer
Verticordia pennigera Endl.
Verticordia plumosa (Desf.) Druce

ONAGRACEAE
Epilobium billardierianum ser. ssp. *billardierum*
Epilobium billardierianum ssp. *cinereum* (A. Rich.) Raven et Englehorn

Epilobium billardierianum ssp. *intermedium* Raven et Englehorn
* *Epilobium ciliatum* Raf.
Epilobium hirtigerum Cunn.
* *Oenothera glazioviana* Micheli
* *Oenothera stricta* Ledeb. ex Link

HALORAGACEAE
Glischrocaryon aureum (Lindley) Orch. var. *aureum*
Glischrocaryon aureum var. *angustifolium* (Nees) Orch.
Glischrocaryon roei Endl.
Gonocarpus benthamii Orch.
Gonocarpus diffusus (Diels) Orch.
(E) *Gonocarpus hexandrus* (F. Muell.) Orch. ssp. *hexandrus*
(E) *Gonocarpus hexandrus* ssp. *serratus* (Schindl.) Orch.
Gonocarpus nodulosus Nees
Gonocarpus paniculatus (R. Br. ex Benth.) Orch.
Gonocarpus simplex (R. Br. ex Britten) Orch.
Haloragis acutangula F. Muell. forma *occidentalis* Orch.
Haloragis brownii (J.D. Hook.) Schindler
Halorhagis digyna Labill
Haloragodendron racemosum (Labill.) Orch.
Meziella trifida (Nees) Schindler
* *Myriophyllum aquaticum* (Vell. Conc.) Verdc.
Myriophyllum crispatum Orch.
Myriophyllum drummondii Benth.
Myriophyllum salsugineum Orch.

APIACEAE
(E) *Actinotus* sp. nov.
Actinotus omnifertilis (F. Muell.) Benth.
Actinotus glomeratus Benth.
* *Centella asiatica* (L) Urban
* *Conium maculatum* L.
* *Daucus carrota* L.
Daucus glochidiatus (Labill.) Fischer
Eryngium pinnatifidum Bunge
* *Foeniculum vulgare* Miller
Homalosciadium homalocarpum (F. Muell.) H. Eichler
* *Ammi majus* L.
Apium annuum P.S. Short
Apium prostratum Labill. ex Vent. ssp. *prostratum*
Apium prostratum var. *filiforme* (A. Rich.) Kirk
Apium prostratum ssp. *phillipii* Keighery
Hydrocotyle alata A. Rich.
Hydrocotyle blepharocarpa F. Muell.
Hydrocotyle callicarpa Bunge
Hydrocotyle diantha DC.
Hydrocotyle hirta R. Br. ex A. Rich.
Hydrocotyle hispidula Bunge var. *hispidula*.
(E) *Hydrocotyle hispidula* var. *tenella* Benth.
Hydrocotyle medicaginoides Turçz.
Hydrocotyle pilifera Turcz. var. *glabrata* Benth.
Hydrocotyle plebeja A. Rich.
Hydrocotyle scutellifera Benth.
Hydrocotyle tetragonocarpa Bunge
Hydrocotyle sp. Hamelin Bay
Pentapeltis peltigera (Hook.) Bunge
Pentapeltis silvatica (Diels) Domin
Platysace compressa (Labill.) Norman
Platysace filiformis (Bunge) Norman
Platysace haplosciadia (Benth) Norman
Platysace ramosissima (Benth.) Norman
Platysace pendula (Benth.) Norman
Platysace tenuissima (Benth.) Norman
Schoenolaena juncea Bunge
Schoenolaena tenuior Bunge
Sirium latifolium L.
Trachymene anisocarpa (Turcz.) B.L. Burtt
Trachymene coerulea R.A. Graham
Trachymene cyanopetala (F.Muell.) Benth.
Trachymene ornata (Endl.) Druce
Trachymene pilosa Smith
Xanthosia atkinsoniana F. Muell.
Xanthosia candida (Benth.) Steudel
Xanthosia hederifolia Benth.
Xanthosia huegelii (Benth.) Steudel
Xanthosia pusilla Bunge
Xanthosia rotundifolia DC.
(E) Gen. nov. sp. nov. (Shannon)

EPACRIDACEAE
Actrotriche cordata (Labill.) R. Br.
Actrotriche depressa R. Br.
(E) *Andersonia auriculata* L. Watson
(E) *Andersonia barbata* L. Watson
Andersonia caerulea R. Br.
Andersonia involucrata Sonder
Andersonia lehmanniana Sonder ssp. *lehmanniana*
Andersonia longifolia (Benth.) L. Watson
Andersonia micrantha R. Br.
Andersonia simplex (Stschegl.) Druce
Andersonia sprengeloides R. Br.
Andersonia sp. nov.
Astroloma baxteri DC.
Astroloma ciliatum (Lindley) Druce
Astroloma drummondii Sonder
Astroloma epacridis (DC.) Druce
Astroloma pallidum R. Br.
Astroloma prostratum R. Br.
Brachyloma concolor (F. Muell.) C. Gardner
Brachyloma preissii Sonder
Conostephium preissii Sonder
Cosmelia rubra R. Br.
(E) *Leucopogon* aff. *tenuis* DC.
Leucopogon alternifolius R. Br.
Leucopogon assimilis R. Br.
Leucopogon australis R. Br.
Leucopogon bracteolaris Benth.
Leucopogon capitellatus DC.
Leucopogon carinatus R. Br.
Leucopogon cinereus E. Pritzel
Leucopogon concinnus Benth.
Leucopogon conostephioides DC.
Leucopogon cordatus Sonder
Leucopogon cucullatus R. Br.
(E) *Leucopogon denticulatus* W. Fitzg.
Leucopogon distans R. Br.
Leucopogon elatior Sonder

Leucopogon flavescens Sonder
(E) *Leucopogon gilbertii* Stschegl.
Leucopogon aff. *gilbertti* Stschegl.
Leucopogon glabellus R. Br.
Leucopogon gracilis R. Br.
Leucopogon gracillimus DC.
Leucopogon hirsutus Sonder
Leucopogon kingianus (F. Muell.) C. Gardner
Leucopogon multiflorus R. Br.
Leucopogon nutans E. Pritzel
Leucopogon obovatus (Labill.) R. Br.
Leucopogon oppositifolius Sonder
Leucopogon ovalifolius Sonder
Leucopogon oxycedrus Sonder
Leucopogon parviflorus (Andrews) Lindley
Leucopogon pendulus R. Br.
Leucopogon polymorphus Sonder
(E) *Leucopogon polystachyus* R. Br.
Leucopogon propinquus R. Br.
Leucopogon pulchellus Sonder
Leucopogon racemulosus DC.
Leucopogon reflexus R. Br.
Leucopogon sprengelioides Sonder
Leucopogon striatus R. Br.
Leucopogon strictus Benth.
Leucopogon unilateralis Stschegl.
Leucopogon verticillatus R. Br.
Lysinema ciliatum R. Br.
Lysinema sp. aff. *ciliatum*
Lysinema conspicuum R. Br.
Lysinema fimbriatum F. Muell.
Lysinema lasianthum R. Br.
Monotoca tamariscina F. Muell.
Needhamiella pumilio (R. Br.) L. Watson
Oligarrhena micrantha R. Br.
Sphenotoma capitatum (R. Br.) Lindley
Sphenotoma gracile (R. Br.) Sweet
Sphenotoma parviflorum F. Muell.
Sphenotoma squarrosum (R. Br.) Don
Styphelia tenuiflora Lindley

PRIMULACEAE
* *Anagallis arvensis* L. var. *arvensis*
* *Anagallis arvensis* var. *caerulea* Gouan.
Samolus junceus R. Br.
Samolus repens (Forster et G. Forster) Pers.
Samolus valerandi L.

LOGANIACEAE
Logania buxifolia F. Muell.
Logania campanulata R. Br.
Logania fasciculata R. Br.
Logania serpyllifolia R. Br.
Logania aff. *serpyllifolia* (G.K.)
Logania vaginalis (Labill.) F. Muell.
Mitrasacme paradoxa R. Br.
Mitrasacme sp. (Annels 2706)

GENTIANACEAE
* *Centaurium erythraea* Rafn.
* *Centaurium spicatum* (L.) Fritsch
Sebaea ovata (Labill.) R. Br.

MENYANTHACEAE
Villarsia albiflora F. Muell.
Villarsia capitata Nees
Villarsia lasiosperma F. Muell.
Villarsia latifolia Benth.
Villarsia parnassifolia (Labill.) R. Br.
Villarsia submersa H.I. Ashton
Villarsia violifolia F. Muell.

APOCYNACEAE
* *Vinca major* L.

ASCLEPIADACEAE
* *Gomphocarpus fruticosus* (L.) W.T. Aiton

CONVOLVULACEAE
* *Calystegia soldanella* R. Br.
Ipomaea indica (Burrman) merr
Dichondra repens Forster and G. Forster
Ipomaea malica (Burrman) merv.
Wilsonia backhousei J.D. Hook.
Wilsonia humilis R. Br.

CUSCUTACEAE
Cuscuta australis R. Br.

BORAGINACEAE
* *Borago officinalis* L.
* *Echium plantagineum* L.
Myosotis australis R. Br.

VERBENACEAE
* *Verbena bonariensis* L.

CHLOANTHACEAE
Pityrodia bartlingii (Lehm.) Benth.

LAMIACEAE
Hemiandra pungens R. Br. var. *pungens*
Hemigenia incana (Lindley) Benth.
Hemigenia microphylla Benth.
Hemigenia podalyrina F. Muell.
Hemigenia sericea Benth.
* *Mentha aquatica* L.
Mentha piperita L.
* *Mentha pulegium* L.
* *Mentha spicata* L.
* *Mentha suaveolens* Ehrh.
Microcorys aff. *obovata* Benth.
* *Prunella vulgaris* L.
* *Salvia verbenacea* L.
Westringia dampieri R. Br.

SOLANACEAE
Anthocercis littorea Labill.
Anthocercis viscosa R. Br. ssp. *viscosa*
* *Datura stramonium* L.
* *Lycium ferocissimum* Miers
* *Nicandra physalodes* (L.) P. Gaertner
* *Physalis peruviana* L.

* *Solanum laciniatum* Aiton
* *Solanum nigrum* L.
Solanum symonii H. Eichler

SCROPHULARIACEAE
* *Bellardia trixago* (L.) All.
* *Dischisma arenarium* E. Meyer
Euphrasia collina R. Br. ssp. *tetragona* (R. Br.) Barker
Euphrasia scabra R. Br.
Glossostigma drummondii Benth.
Gratiola peruviana L.
Morgania floribunda Benth
* *Parentucellia latifolia* (L.) Caruel
* *Parentucellia viscosa* (L.) Caruel
* *Verbascum virgatum* Stokes
* *Veronica arvensis* L.
Veronica calycina R. Br.
Veronica distans R. Br.
Veronica plebeia R. Br.

OROBANCHACEAE
* *Orobanche minor* Smith

LENTIBULARIACEAE
Utricularia menziesii R. Br.
Utricularia multifida R. Br.
Utricularia tenella R. Br.
Utricularia simplex R. Br.
Utricularia violacea R. Br.
Utricularia volubilis R. Br.

MYOPORACEAE
Myoporum apiculatum A. DC.
Myoporum gracile Bartling
Myoporum insulare R. Br.
Myoporum oppositifolium R. Br.
Myoporum tetrandrum (Labill.) Domin

PLANTAGINACEAE
Plantago debilis R. Br.
* *Plantago lanceolata* L.
* *Plantago major* L.

RUBIACEAE
* *Galium murale* (L.) All.
Opercularia echinocephala Benth.
Opercularia hispidula Endl.
Opercularia vaginata Labill.
Opercularia volubilis R. Br. ex Benth.

VALERIANACEAE
* *Centranthus ruber* (L.) DC.

CUCURBITACEAE
* *Cucumis myriocarpus* Naudin

CAMPANULACEAE
Wahlenbergia communis Carolin
Wahlenbergia gracilenta Loth.
Wahlenbergia litticola P.J. Smith
Wahlenbergia multicaulis Benth
Wahlenbergia preissii Vriese
Wahlenbergia simplicicaulis Vriese

LOBELIACEAE
Grammatotheca bergiana (Cham.) C. Presl
Isotoma hypocrateriformis (R. Br.) Druce
Isotoma scapigera (R. Br.) Don
Lobelia alata Labill.
Lobelia heterophylla Labill.
Lobelia gibbosa Labill.
Lobelia rariflora F. Winner
Lobelia rhombifolia Vriese
Lobelia rhytidosperma Benth.
Lobelia tenuior R. Br.
* *Monopsis simplex* (L.) F. Wimmer

GOODENIACEAE
Anthotium humile R. Br.
Dampiera fasiculata R. Br.
Dampiera hederacea R. Br.
(E) *Dampiera heteroptera* Rajput et Carolin
Dampiera leptoclada Benth.
Dampiera linearis R. Br.
Dampiera pedunculata Rajput et Carolin
Dampiera trigona Vriese
Diaspasis filifolia R. Br.
Goodenia caerulea R. Br.
Goodenia concinna Benth.
Goodenia eatoniana F. Muell.
Goodenia filiformis R. Br. var. *filiformis*
Goodenia filiformis var. *pulchella* Benth.
Goodenia incana R. Br.
Goodenia laytoniana Benth.
Goodenia leptoclada Benth.
Goodenia pulchella Benth.
Goodenia sepalosa F. Muell. ex Benth.
var. *glandulosa* F. Muell.
Goodenia tenella R. Br.
Lechenaultia biloba Lindley
Lechenaultia expansa R. Br.
Lechenaultia floribunda Benth.
Lechenaultia formosa R. Br.
Lechenaultia tubiflora R. Br.
(E) *Scaevola attenuata* R. Br.
Scaevola auriculata Benth.
Scaevola calliptera Benth.
Scaevola lanceolata Benth
Scaevola crassifolia Labill.
Scaevola glandulifera DC.
Scaevola globulifera Labill.
Scaevola longifolia Vriese
Scaevola microphylla Benth.
Scaevola nitida R. Br.
Scaevola pilosa Benth.
Scaevola striata R. Br.
Scaevola thesioides Benth.
Selliera radicans Cav.
(E) *Velleia macrophylla* (Lindley) Benth.
Velleia trinervis Labill.

STYLIDIACEAE
Levenhookia dubia Sonder

Levenhookia leptantha Benth.
Levenhookia pauciflora Benth.
Levenhookia preissii (Sonder) F. Muell.
Levenhookia pusilla R. Br.
Stylidium adnatum R. Br.
Stylidium affine Sonder
Stylidium amoenum R. Br.
Stylidium assimile R. Br.
Stylidium barleei F. Muell.
Stylidium beaugleholei J.H. Willis
Stylidium breviscapum R. Br.
Stylidium brunonianum ssp. *brunonianum*
Stylidium brunonianum Benth. ssp. *minor* Carlq.
Stylidium bulbiferum Benth
Stylidium caespitosum R. Br.
Stylidium calcaratum R. Br.
Stylidium aff. *calcaratum* R. Br.
Stylidium canaliculatum Lindley
Stylidium carnosum Benth.
Stylidium ciliatum Lindley
Stylidium corymbosum R. Br.
Stylidium crassifolium R. Br.
Stylidium despectum R. Br.
Stylidium dichotomum DC.
Stylidium ecorne (F. Muell. ex Erickson et J.H. Willis) P.G. Farrell et S.H.James
Stylidium exoglossum R. Erickson et J.H. Willis
Stylidium falcatum R. Br.
Stylidium fasciculatum R. Br.
Stylidium glaucum Labill. ssp. *glaucum*
(E) *Stylidium glaucum* ssp. *angustifolium* Carlq.
Stylidium guttatum R. Br.
Stylidium hirsutum R. Br.
Stylidium imbricatum Benth.
Stylidium inundatum R. Br.
Stylidium junceum R. Br. ssp. *junceum*
Stylidium junceum ssp. *brevis* (E. Pritzel) Carlq.
(E) *Stylidium laciniatum* C. Gardner
Stylidium lepidium F. Muell. ex Benth.
Stylidium luteum R. Br. ssp. *luteum*
Stylidium luteum ssp. *glaucifolium* Carlq.
Stylidium periscelianthum R. Erickson et J.H. Willis
Stylidium perpusillum J.D. Hook.
Stylidium petiolare Sonder
Stylidium piliferum R. Br.
Stylidium preissii (Sonder) F. Muell
(E) *Stylidium pritzelianum* Mildbr.
Stylidium pulchellum Sonder
(E) *Stylidium pygmaeum* R. Br.
Stylidium repens R. Br.
Stylidium rhynchocarpum Sonder
Stylidium rupestre Sonder
Stylidium scandens R. Br.
(E) *Stylidium* aff. *scandens* R. Br.
Stylidium schoenoides DC.
Stylidium spathulatum R. Br. ssp. *spathulatum*
(?E) *Stylidium spathulatum* ssp. *acuminatum* Carlq.
Stylidium spinulosum R. Br.
Stylidium squamosotuberosum Carlq.
Stylidium uniflorum Sonder
Stylidium violaceum R. Br.

ASTERACEAE
Actites megalocarpa (J.D. Hook.) N.S. Lander
Angianthus preissianus (Steetz) Benth.
* *Arctotheca calendula* (L.) Levyns
* *Arctotheca populifolia* (P. Bergius) Norlindh
Asteridea nivea (Steetz) G. Kroner
Astridea gracilis A. Gray.
Asteridea pulverulenta Lindley
Berkheya rigida (Thunb.) Ewart, J. White et B. Rees
Blennospora drummondii A. Gray.
Brachycome ciliaris (Labill.) Less.
Brachycome exilis Sander
Brachycome iberidifolia Benth.
Calocephalus brownii (Cass.) F. Muell.
* *Carduus pycnocephalus* L.
Carduus tenuiflorus Curt
* *Centaurea melitensis* L.
Centipeda cunninghamii (DC.) A. Braun et Asch.
* *Chrysanthemum segetum* L.
Chrysocoryne pusilla (Benth.) Endl.
* *Cirsium arvense* (L.) Scop.
* *Cirsium vulgare* (Savi) Ten.
* *Conyza albida* Willd. ex Sprengel
* *Conyza bonariensis* (L.) Cronq.
* *Conyza parva* Cronq.
* *Coreopsis grandiflora* Hogg ex Sweet
Cotula australis (Sieber ex Sprengel) J.D. Hook.
Cotula coronopifolia L.
Cotula cotuloides (Steetz.) Druce
Cotula drummondii Benth.
* *Cotula turbinata* L.
Craspedia pleiocephala F. Muell.
* *Crepis foetida* L.
* *Cynara cardunculus* L.
* *Dittrichia graveolens* (L.) Greuter
* *Dittrichia viscosa* (L.) Greuter
Gnaphalium gymnocephalum DC.
* *Gnaphalium pensylvanicum* Willd.
Gnaphalium indutum J.D. Hook
Gnaphalium sphaericum Willd.
* *Hedypnois rhagadioloides* (L.) F.W. Schmidt
Helichrysum cordatum DC.
Helichrysum macranthum Benth.
Helichrysum obtusifolium F. Muell. et Sonder
Helichrysum ramosum DC.
Helipterum pygmaeum (DC) Benth
Hyalosperma cotula (Benth.) P.G. Wilson
Hyalosperma pusillum (Turcz) P.G. Wilson
Hyalosperma simplex (Steetz) P.G. Wilson
* *Hypochaeris glabra* L.
Ixiolaena viscosa Benth.
Lagenifera huegelii Benth.
Leptorhynchos scabrus L. Haegi
Millotia myosotidifolia (Benth.) Steetz
Millotia tennuifolia Cass
Olearia axillaris (DC.) F. Muell. ex Benth.
Olearia calcarea F. Muell
Olearia cassiniae Benth.
Olearia ciliata (Benth.) F. Muell.
Olearia elaeophila (DC.) F. Muell. ex Benth.
Olearia paucidentata (Steetz) F. Muell. ex Benth.
Olearia revoluta F. Muell. ex Benth.

Olearia rudis (Benth.) F. Muell. ex Benth.
Osteospermum clandestinum (Lees) Norlindh
* *Pentzia suffruticosa* (L.) Hutch. ex Merxm.
Picris squarrosa Stectz
Pithocarpa corymbulosa Lindley
Pithocarpa melanostigma P. Lewis et Summerh.
Podolepis canescens Cunn. ex DC.
Podolepis gracilis (Lehm.) R.A. Graham
Podolepis lessonii (Cass.) Benth.
Podolepis rugata Labill.
Podotheca angustifolia (Labill.) Less.
* *Pseudognaphalium luteo-album* (L.) Hilliard et B.L. Burtt
Quinetia urvillei Cass.
Rutidosis multiflora (Nees) Robinson
* *Senecio diaschides* D. Drury
* *Senecio elegans* L.
Senecio gilbertii Turcz.
Senecio glomeratus Desf. ex Poiret
Senecio glossanthus (Sonder) Belcher
* *Senecio jacobaea* L.
Senecio lautus G. Forster ex Willd. ssp. *maritimus* Ali.
* *Senecio mikanioides* Otto ex Walp.
Senecio minimus Poiret
Senecio quadridentatus Labill.
Senecio ramosissimus DC.
* *Senecio vulgaris* L.
* *Sigesbeckia orientalis* L.
Siloxerus filifolius (Benth.) Ostenf.
Siloxerus humifusus Labill.
* *Silybum marianum* (L.) Gaertner
* *Soliva pterosperma* (A.L. Juss.) Less.
* *Sonchus asper* Hill
* *Sonchus oleraceus* L.
Trichocline spathulata (Cunn. ex DC.) J.H. Willis
* *Ursinia anthemoides* (L.) Poiret
* *Ursinia speciosa* DC.
* *Vellereophyton dealbatum* (Thunb.) Hilliard et B.L. Burtt.
Waitzia citrina (Benth.) Steetz
Waitzia paniculata (Steetz) F.Muell. ex Benth.
Waitzia suaveolens (Benth.) Druce
* *Xanthium occidentale* Bertol.

Appendix 2
CONSERVATION RESERVES IN THE SOUTHERN FORESTS AS PROPOSED BY CALM (1987a, 1987b)[a,b]

Name (or locality)	Area (ha)
NATURE RESERVES	
Busselton District	
Whicher	6 370
Chester	330
Paget	1 400
Mowen	820
Kirup District	
Greenbushes	1 400
Nannup District	
East Nannup Road	10
Gingilup Swamp	2 800
Dalgarup	920
Milyeannup	5 560
Manjimup District	
(Wilgarup)	80
Big Brook	90
Bokarup	660
Cobertup	140
Cowerup	270
Donnelly River	60
Galamup	220
Kodjinup	620
Kulunilup	610
Lake Muir	11 400
Mordalup	110
Noobijup	190
Pindicup	370
(Smith Br.) (part)	50
Unicup	3 300
(Quindinup)	560
(Palgarrup North)	10
(Wilgarup)	200
Yornup	70
(Dickson)	550
(Strickland)	1 700
(Bolbelup)	2 100
(Blackbutt reserve)	40
Perup	40 200
Pemberton District	
Flat and Sandy Is.	20
Smith Br (part)	50
(Eastbrook Road)	20
(East Brook)	60
(Jane Block)	10
Walpole District	
L. Eyrie	20
Blue Gum Creek	100
Chatham Is.	80
Kwornicup L.	250
Mt. Shadford	80
(Pardelup)	690
Tootanellup	980
(Onerup Lagoon)	50
(L. Barnes Rd)	280
(Scotsdale Rd)	20
(Redmond Rd)	60
(McIntosh Rd)	30
(McLean Rd)	10
(Randell Rd)	460
Wambellup	510
(Chorkerup)	30
Sleeman Ck	410
(Gum Link Rd)	600
Thames	470
(Kordabup)	480
CONSERVATION PARKS	
Blackwood	3 230
(Bridgetown)	50
(Ellis Creek)	130
Mullalyup	2 200
Rapids	2 500
St John Brook	3 400
Boorara	610
Dombakup	110
Muirillup	160
One Tree Bridge	640
(Jervik Park)	20
(Jardee)	10
NATIONAL PARKS	
Leeuwin-Naturaliste	19 700
Scott	3 300
D'Entrecasteaux	123 200
(Mt. Lindesay-Sheepwash)	27 000
Pemberton	5 900
Shannon	54 000
Walpole-Nornalup	18 800
Wattle-Soho, Mt. Frankland	30 300

NOTE - Areas under 1000 ha are to nearest 10 ha
Areas over 1000 ha are to nearest 100 ha

[a]CALM (1987a). *Regional Management Plan 1987-1997, Central Forest Region.* Department of Conservation and Land Management, Como, W.A.

[b]CALM (1987b). *Regional Management Plan 1987-1997, Southern Forest Region.* Department of Conservation and Land Management, Como, W.A.

Appendix 3
FAUNA OF THE SOUTHERN FORESTS

Appendix 3a MAMMALS RECORDED IN MAJOR NATIONAL PARKS AND NATURE RESERVES

Legend

1. D'Entrecasteaux (NP)
2. Perup (NR)
3. Leeuwin-Naturaliste (NP)
4. St Johns Brook (CP) - Whicher Range (NR)
5. Milyeannup (NR)
6. Mt. Frankland (NP)
7. Shannon (NP)
8. Warren (NP)
9. Mt. Lindesay (NP)
10. Walpole-Nornalup (NP)
11. Other[a]

NP = National Parks
NR = Nature Reserves
CP = Conservation Parks

	National Parks or Reserves										
	1	2	3	4	5	6	7	8	9	10	11
Order Diprotodonta **Family Macropodidae**											
Western Grey Kangaroo *(Macropus fuliginosus)*	X	X	X	X	X	X	X	X	X	X	X
Western Brush Wallaby *(Macropus irma)*		X	X	X	X	X			X		X
Tammar Wallaby *(Macropus eugenii)*		X									
Quokka *(Setonix brachyurus)*	X		X	X		X	X	X		X	X
Brush-tailed Bettong or Woylie *(Bettongia penicillata)*		X									X
Family Phalangeridae											
Common Brushtail possum *(Trichosurus vulpecula)*		X	X	X	X			X	X		X
Family Petauridae											
Common Ringtail possum *(Pseudocheirus peregrinus)*		X	X					X	X		X
Family Burramyidae											
Western Pygmy possum *(Cercartetus concinnus)*	X	X		X	X			X	X		
Family Tarsipedidae											
Honey possum *(Tarsipes rostratus)*	X			X	X	X	X		X	X	X
Order Polyprotodonta **Family Peramelidae**											
Southern Brown Bandicoot *(Isoodon obesulus)*	X	X	X	X		X	X	X	X	X	X

NOTE: All species of vertebrates listed are recent records having been recorded in the area during the last 20 years.
[a] Other - includes species recorded outside the forested areas in farmland and pine plantations.

	1	2	3	4	5	6	7	8	9	10	11
Family Dasyuridae											
Western Quoll or Chuditch (*Dasyurus geoffroii*)		X		X							
Brush-tailed Phascogale (*Phascogale tapoatafa*)		X						X			X
Yellow-footed Antechinus or Mardo (*Antechinus flavipes*)	X	X			X	X		X		X	X
Grey-bellied Dunnart (*Sminthopsis griseoventer*)	X	X	X	X	X	X	X	X	X	X	X
Family Myrmecobiidae											
Numbat (*Myrmecobius fasciatus*)		X									X
Order Rodentia											
Family Muridae											
Bush Rat (*Rattus fuscipes*)	X	X	X	X	X	X	X	X	X	X	X
Water-rat (*Hydromys chrysogaster*)	X	X		X		X	X	X			
Order Chiroptera											
Family Vespertilionidae											
Greater Long-eared Bat (*Nyctophilus major*)	X	X				X					
Gould's Long-eared Bat (*Nyctophilus gouldi*)											
Lesser Long-eared Bat (*Nyctophilus geoffroyi*)	X	X							X		
Gould's Wattled Bat (*Chalinolobus gouldii*)		X	X	X					X		
Chocolate Wattled Bat (*Chalinolobus morio*)		X				X	X				
King River Eptesicus (*Eptesicus regulus*)	X	X	X	X		X	X		X		
Falsistrellus mackenziei	X	X	X	X		X		X	X		
White-Striped Mastiff-bat (*Tadarida australis*)											X
Little Mastiff-bat (*Mormopeterus planiceps*)		X									X
Order Monotremata											
Family Tachyglossidae											
Echidna (*Tachyglossus aculeatus*)		X									X

	1	2	3	4	5	6	7	8	9	10	11
EXOTIC MAMMALS											
Order Carnivora **Family Felidae**											
Cat *(Felis catus)*	X	X	X	X	X	X	X	X	X	X	X
Family Canidae											
Dingo *(Canis familiaris dingo)*	X	X		X		X					
Fox *(Vulpes vulpes)*	X	X	X	X	X	X	X	X	X	X	X
Order Artiodactyla **Family Suidae**											
Pig *(Sus scrofa)*					X						
Family Bovidae											
Goat *(Capra hircus)*								X			X
Order Perissodactyla **Family Equidae**											
Horse *(Equus caballus)*				X							X
Order Rodentia **Family Muridae**											
House Mouse *(Mus musculus)*	X	X	X	X	X	X	X	X	X	X	X
Black Rat *(Rattus rattus)*	X		X	X	X	X	X	X	X	X	X
Order Lagomorpha **Family Leporidae**											
Rabbit *(Oryctolagus cuniculus)*	X	X	X	X		X	X	X	X	X	X

Appendix 3b BIRDS RECORDED IN MAJOR NATIONAL PARKS AND NATURE RESERVES

LEGEND

1. D'Entrecasteaux (NP)
2. Perup (NR)
3. Leeuwin-Naturaliste (NP)
4. St Johns Brook (CP)- Whicher Range (NR)
5. Milyeannup (NR)
6. Mt. Frankland (NP)
7. Shannon (NP)
8. Warren (NP)
9. Mt. Lindesay (NP)
10. Walpole-Nornalup (NP)
11. Other[a]

NP = National Parks
NR = Nature Reserves
CP = Conservation Parks

	NATIONAL PARKS OR RESERVES										
	1	2	3	4	5	6	7	8	9	10	11
Order Struthioniformes **Family Dromaiidae**											
Emu *(Dromaius novaehollandiae)*	X	X	X	X	X	X	X	X	X	X	X
Order Podicipediformes **Family Podicipedidae**											
Hoary-headed Grebe *(Poliocepahalus poliocephalus)*							X				
Australasian Grebe *(Tachybaptus novaehollandiae)*	X	X		X		X					X
Great Crested Grebe *(Podiceps cristatus)*											X
Order Procellariiformes **Family Diomedeidae**											
Huttons Shearwater *(Puffinus huttoni)*	X										
Order Pelecaniformes **Family Pelecanidae**											
Australian Pelican *(Pelecanus conspicillatus)*							X			X	
Family Anhingidae											
Darter *(Anhinga melanogaster)*	X	X		X						X	X
Family Phalacrocoracidae											
Great cormorant *(Phalacrocorax carbo)*	X						X			X	
Pied cormorant *(Phalacrocorax varius)*	X		X				X				
Little Black Cormorant *(Phalacrocorax sulcirostris)*		X		X		X					X
Little Pied Cormorant *(Phalacrocorax melanoleucus)*	X	X		X		X	X		X	X	X

NOTE: All species of vertebrates listed are recent records having been recorded in the area during the last 20 years.
[a] Other - includes species recorded outside the forested areas in farmland and pine plantations.

	1	2	3	4	5	6	7	8	9	10	11
Order Ardeiformes											
Family Ardeidae											
Pacific Heron		X				X	X		X	X	X
(Ardea pacifica)											
White-faced Heron	X	X		X		X	X	X	X	X	X
(Ardea novaehollandiae)											
Great Egret	X			X							X
(Egretta alba)											
Rufous Night Heron	X	X				X					X
(Nycticora caladonicus)											
Black Bittern		X				X					
(Dupetor flavicollis)											
Australasian Bittern	X			X							
(Botaurus poiciloptilus)											
Family Plataleidae											
Sacred Ibis	X										X
(Threskiornis aethiopica)											
Straw-necked Ibis	X	X	X								X
(Threskiornis spinicollis)											
Yellow-billed Spoonbill											X
(Platula flavipes)											
Order Anseriformes											
Family anatidae											
Black Swan		X					X				X
(Cygnus atratus)											
Freckled Duck	X										
(Stictonetta naevosa)											
Australian Shelduck		X		X		X	X				X
(Tadorna tadornoides)											
Pacific Black Duck	X	X	X	X	X	X	X		X	X	X
(Anas superciliosa)											
Grey Teal	X	X									X
(Anas gibberifrons)											
Australasian Shoveller	X										
(Anas rhynchotis)											
Maned Duck		X		X		X				X	X
(Chenonetta jubata)											
Blue-billed Duck	X										
(Oxyura australis)											
Musk Duck	X	X		X		X	X				X
(Biziura lobata)											
Order Accipitriformes											
Family Pandionidae											
Osprey	X						X				
(Pandion haliaetus)											
Family Accipitridae											
Black-shouldered Kite										X	X
(Elanus notatus)											
Square-tailed Kite	X										
(Lophoictinia isura)											

	1	2	3	4	5	6	7	8	9	10	11
Whistling Kite *(Helistur sphenurus)*	X	X	X	X		X		X	X		X
Brown Goshawk *(Accipter fasciatus)*	X	X		X		X			X		X
Collared Sparrowhawk *(Accipiter cirrhocephalus)*					X			X			X
White-bellied Sea Eagle *(Haliaeetus leucogaster)*	X						X				
Wedge-tailed Eagle *(Aquila audax)*	X	X		X		X	X	X	X	X	X
Little Eagle *(Hieraaetus morphnoides)*		X		X		X					
Spotted Harrier *(Circus assimilis)*	X										
Marsh Harrier *(Circus aeruginosus)*							X			X	
Family Falconidae											
Peregrine Falcon *(Falco peregrinus)*	X					X					
Australian Hobby *(Falco longipennis)*	X	X									X
Brown Falcon *(Falco berigora)*	X	X	X	X	X	X	X		X	X	X
Australian Kestrel *(Falco cenchroides)*	X	X	X			X			X	X	X
Order Galliformes **Family Megapodiidae**											
Malleefowl *(Leipoa ocellata)*						X					
Family Phasianidae											
Stubble Quail *(Coturni novaezelandiae)*						X	X		X		X
Brown Quail *(Coturni australis)*											X
Order Gruiformes **Family Turnicidae**											
Painted Button-quail *(Turni varia)*		X									
Family Rallidae											
Buff-banded Rail *(Rallus philippensis)*											X
Baillon's Crake *(Porzona pusilla)*											X
Australian Crake *(Porzona fluminea)*											X
Spotless Crake *(Porzana tabuensis)*	X			X	X			X			
Dusky Moorhen *(Gallinula tenebrosa)*											X

	1	2	3	4	5	6	7	8	9	10	11
Purple Swamphen	X						X	X			X
(Porphyrio porphyrio)											
Eurasian Coot	X	X					X				X
(Fulica atra)											
Order Charadriiformes											
Family Charadriidae											
Banded Lapwing	X	X									
(Vanellus tricolor)											
Red-capped Plover	X						X				X
(Charadrius ruficapillus)											
Black-fronted Plover		X									X
(Charadrius melanops)											
Family Scolopacidae											
Greenshank											X
(Tringa nebularia)											
Curlew Sandpiper	X						X				
(Calidris ferruginea)											
Family Recurvirostridae											
Black-winged stilt											X
(Himantopus himantopus)											
Family Laridae											
Silver Gull	X		X				X				X
(Larus novaehollandiae)											
Pacific Gull	X						X				X
(Larus pacificus)											
Caspian Tern	X		X				X				X
(Hydroprogne caspia)											
Crested Tern	X		X				X				X
(Sterna bergii)											
Order Columbiformes											
Family Columbidae											
Common Bronzewing	X	X	X	X							X
(Phaps chalcoptera)											
Brush Bronzewing	X	X	X	X	X	X	X	X	X	X	X
(Phaps elegans)											
Order Psittaciformes											
Family Cacatuidae											
Red-tailed Black Cockatoo	X	X		X		X			X		X
(Calyptorhyncus magnificus)											
White-tailed Black Cockatoo	X	X	X	X	X	X	X	X	X	X	X
(Calyptorhynchus baudinii)											
Family Loriidae											
Purple-crowned Lorikeet	X	X				X	X	X	X	X	X
(Glossopsitta porphyocephala)											

	1	2	3	4	5	6	7	8	9	10	11
Family Platycercidae											
Red-capped Parrot *(Purpureicephalus spurius)*	X	X		X	X	X	X	X	X		X
Western Rosella *(Platycercus icterotis)*	X	X	X	X	X	X	X	X	X	X	X
Port Lincoln Ringneck *(Barnardius zonarius)*	X	X	X	X	X	X	X	X	X	X	X
Elegant Parrot *(Neophema elegans)*	X	X		X					X		X
Rock Parrot *(Neophema petrophila)*	X		X								X
Order Cuculiformes **Family Cuculidae**											
Pallid Cuckoo *(Cuculus pallidus)*	X	X		X	X	X	X	X	X	X	X
Fan-tailed Cuckoo *(Cuculus pyrrhophanus)*	X	X	X		X	X	X	X		X	X
Horsefield's Bronze Cuckoo *(Chrysococcyx basilis)*	X		X		X	X					
Shining Bronze-Cuckoo *(Chrysococcyx lucidus)*		X		X	X	X	X	X		X	X
Order Strigiformes **Family Strigidae**											
Southern Boobook *(Ninox novaeseelandiae)*	X	X	X	X		X		X			X
Family Tytonidae											
Barn Owl *(Tyto alba)*		X	X								X
Masked Owl *(Tyto novaehollandiae)*						X					
Order Caprimulgiformes **Family Podargidae**											
Tawny Frogmouth *(Podargus strigoides)*	X	X	X	X	X	X	X	X	X		X
Family Aegothelidae											
Australian Owlet-nightjar *(Aegotheles cristatus)*	X	X		X		X	X	X	X		X
Family Caprimulgidae											
Spotted Nightjar *(Caprimulgus guttatus)*						X					
Order Coraciiformes **Family Alcedinidae**											
Sacred Kingfisher *(Halycon sancto)*	X	X	X	X		X	X	X		X	X

	1	2	3	4	5	6	7	8	9	10	11
Family Meropidae											
Rainbow Bee-eater *(Merops ornatus)*				X							X
Order Passriformes **Family Hirundinidae**											
Welcome Swallow *(Hirundo neoxena)*	X	X		X			X		X		X
Tree Martin *(Cereopis nigricans)*	X	X	X	X	X	X	X	X		X	X
Family Motacillidae											
Richard's Pipit *(Anthus novaeseelandiae)*	X	X	X	X		X	X	X	X	X	X
Family Campephagidae											
Black-faced Cuckoo-shrike *(Coracina novaehollandiae)*	X	X	X	X	X	X	X	X	X	X	X
White-winged Triller *(Lalage Sueurii)*		X									X
Family Muscicapidae											
Scarlet Robin *(Petroica multicolor)*	X	X	X	X	X	X	X	X	X	X	X
Red-capped Robin *(Petroica goodenovii)*											X
White-breasted Robin *(Eopsaltria georgiana)*	X	X	X	X	X	X	X	X	X	X	X
Western Yellow Robin *(Eopsaltria griseogularis)*	X	X		X	X			X	X		X
Crested Shrike-tit *(Falcunculus frontatus)*	X	X						X			X
Golden Whistler *(Pachycephala pectoralis)*	X	X	X	X	X	X	X	X	X	X	X
Rufous Whistler *(Pachycephala rufiventris)*		X	X						X		X
Grey Shrike-thrush *(Colluricincla harmonica)*	X	X		X	X	X		X	X	X	X
Restless Flycatcher *(Myiagra inquieta)*	X	X	X			X	X		X		X
Grey Fantail *(Rhipidura fuliginosa)*	X	X	X	X	X	X	X	X	X		X
Willie Wagtail *(Rhipidura leucophrys)*	X	X	X	X					X		X
Family Epthianuridae											
White-fronted Chat *(Epthianura albifrons)*											X
Family Timaliidae											
White-browed Babbler *(Pomatostomus superciliosus)*	X			X		X	X	X		X	X

	1	2	3	4	5	6	7	8	9	10	11
Family Sylviidae											
Clamorous Reed-Warbler *(Acrocephalus stentoreus)*	X										X
Little Grassbird *(Megalurus gramineus)*	X										
Rufous Songlark *(Cinclorhamphus mathewsi)*											X
Brown Songlark *(Cinclorhampus cruralis)*											X
Family Maluridae											
Splendid Fairy-wren *(Malurus splendens)*	X	X	X	X	X	X	X	X	X	X	X
Red-winged Fairy-wren *(Malurus elegans)*	X	X	X	X	X	X	X	X			X
Southern Emu-wren *(Stipiturus malachurus)*	X		X	X	X	X				X	
Family Acanthizidae											
White-browed Scrub-wren *(Smicrornis frontalis)*	X	X	X	X	X	X	X	X	X	X	X
Weebill *(Smicrornis brevirostris)*		X		X	X	X					
Western Gerygone *(Gerygone fusca)*	X	X		X	X	X		X		X	X
Inland Thornbill *(Acanthiza apicalis)*	X	X	X	X	X	X	X	X	X	X	X
Western Thornbill *(Acanthiza inornata)*	X	X		X	X	X	X		X	X	X
Yellow-rumped Thornbill *(Acanthiza chrysorrhoa)*	X	X		X		X	X		X		X
Family Neosittidae											
Varied Sitella *(Daphhoenositta chrysoptera)*	X	X	X	X	X	X	X	X	X	X	X
Family Climacteridae											
Rufous Treecreeper *(Climacteris rufa)*	X	X		X		X	X	X	X	X	X
Family Meliphagidae											
Red Wattlebird *(Anthochaera carunculata)*	X	X	X	X		X	X	X	X	X	X
Little Wattlebird *(Anthochaera chrysoptera)*	X	X		X	X	X	X				X
Yellow-throated Minor *(Manorina flavigula)*	X										
Singing Honeyeater *(Lichenostomus ornatus)*		X									X
White-naped Honeyeater *(Melithreptus lunatus)*	X	X	X	X	X	X	X	X	X	X	X
Brown Honeyeater *(Lichmera indistincta)*	X	X				X			X		X

	1	2	3	4	5	6	7	8	9	10	11
New Holland Honeyeater *(Phylidonyris novaehollandiae)*	X	X	X	X	X	X	X	X	X	X	X
White-cheeked Honeyeater *(Phylidonyris nigra)*	X										
Tawny-crowned Honeyeater *(Phylidonyris melanops)*	X				X	X	X				X
Western Spinebill *(Acanthorhynchus superciliosus)*	X	X		X	X	X	X		X	X	X
Family Pardalotidae											
Spotted Pardalote *(Pardalote punctatus)*	X	X	X	X	X	X	X	X	X		X
Striated Pardalote *(Pardalotus striatus)*	X	X	X	X	X	X		X		X	X
Family Zosteropidae											
Silvereye *(Zosterops lateralis)*	X	X	X	X	X	X	X	X	X	X	X
Family Ploceidae											
Red-eared Fire-tail *(Emblema oculata)*	X		X	X		X	X	X	X	X	X
Family Grallinidae											
Australian Magpie-lark *(Grallina cyanoleuca)*	X	X	X	X			X		X	X	X
Family Artamidae											
Dusky Woodswallow *(Artamus cyanopterus)*	X	X	X	X		X	X	X	X	X	X
Family Cracticidae											
Grey Butcherbird *(Cracticus torquatus)*	X										X
Australian Magpie *(Gymnorhina tibicen)*	X	X	X	X	X	X	X	X	X	X	X
Grey Currawong *(Strepera versicolor)*	X	X	X	X	X	X	X	X	X	X	X
Family Corvidae											
Australian Raven *(Corvus coronoides)*	X	X	X	X	X	X	X	X	X	X	X
EXOTIC BIRDS											
Order Coraciiformes **Family Alcedinidae**											
Laughing Kookaburra *(Dacelo novaeguinea)*	X	X	X	X	X	X	X	X	X	X	X

Appendix 3c REPTILES RECORDED IN MAJOR NATIONAL PARKS AND NATURE RESERVES

Legend

1. D'Entrecasteaux (NP)
2. Perup (NR)
3. Leeuwin-Naturaliste (NP)
4. St Johns Brook (CP)-
   Whicher Range (NR)
5. Milyeannup (NR)
6. Mt. Frankland (NP)
7. Shannon (NP)
8. Warren (NP)
9. Mt. Lindesay (NP)
10. Walpole-Nornalup (NP)
11. Other[a]

NP = National Parks
NR = Nature Reserves
CP = Conservation Parks

	National Parks or Reserves										
	1	2	3	4	5	6	7	8	9	10	11
Order Testudines **Family Cheloniidae**											
Long-necked Tortoise *(Chelodina oblonga)*	X	X					X	X	X	X	X
Order Squamata **Family Gekkonidae**											
Marbled Gecko *(Phyllodactylus marmoratus)*	X	X	X	X		X	X		X		X
Family Pygopodidae											
The Common Scaly-foot *(Pygopus lepidopodus)*	X		X				X				
Fraser's Scale-footed Lizard *(Delma fraseri)*	X	X									
Pretty Worm-lizard *(Aprasia pulchella)*					X						
Family Scincidae											
Mourning Skink *(Egernia luctuosa)*	X			X	X	X					
Bobtail *(Tiliqua rugosa)*	X	X	X	X	X	X					
Smith's Skink *(Egernia napoleonis)*	X	X	X	X	X	X					
King's Skink *(Egernia kingii)*	X	X		X	X	X					
Fry's Skink *(Egernia pulchra pulchra)*	X			X	X	X					
Red-legged Skink *(Ctenotus labillardieri)*	X	X	X	X	X	X					
Ctenotus catenifer				X		X					
Ctenotus impar			X	X							
*Hemiergis initialis initialis*X				X							
Slippery Skink *(Lerista microtis microtis)*		X			X	X					
Sphenomorphus australis	X					X					
Burrowing Skink *(Hemiergis peronii peronii)*	X	X	X	X	X	X					

NOTE: All species of vertebrates listed are recent records having been record(
[a] Other - includes species recorded outside the forested areas in farmland an(

	1	2	3	4	5	6	7	8	9	10	11
New Holland Skink	X	X		X					X		X
(Leiolopisma trilineatum)											
Wood Skink			X	X					X		
(Cryptoblepharus plagiocephalus)											
Sandhill Skink	X			X							X
(Morethia lineoocellata)											
Morethia obscura	X			X							X
Lerista elegans											X
Lerista distinguenda			X	X							X
Grey's Skink			X	X							X
(Menetia greyii)											
Bungurra	X	X	X	X				X	X		X
(Varanus rosenbergi)											
Family Typhlopidae											
Blind Snake		X		X	X						X
(Ramphotyphlops australis)											
Family Boidae											
Carpet Python											x
(Morelia spilota variegata)											
Family Elapidae											
Dugite	X	X	X	X	X	X	X	X	X	X	X
(Pseudonaja affinis affinis)											
Black Tiger Snake	X	X		X		X	X	X	X	X	X
(Notechis ater occidentalis)											
The Black-headed Snake		X			X						
(Unechis gouldii)											
Crowned Snake	X		X		X	X	X		X	X	X
(Drysdalia coronata)											
Short-nosed Snake							X				X
(Notechis minor)											
Square-nosed Snake	X			X	X	X	X		X		
(Rhinoplocephalus bicolor)											
Bardick									X		X
(Echiopsis curta)											

Appendix 3d AMPHIBIANS RECORDED IN MAJOR NATIONAL PARKS AND NATURE RESERVES

LEGEND

1. D'Entrecasteaux (NP)
2. Perup (NR)
3. Leeuwin-Naturaliste (NP)
4. St Johns Brook (CP)- Whicher Range (NR)
5. Milyeannup (NR)
6. Mt. Frankland (NP)
7. Shannon (NP)
8. Warren (NP)
9. Mt. Lindesay (NP)
10. Walpole-Nornalup (NP)
11. Other[a]

NP = National Parks
NR = Nature Reserves
CP = Conservation Parks

	NATIONAL PARKS OR RESERVES										
	1	2	3	4	5	6	7	8	9	10	11
Order Salienta											
Family Hylidae											
Slender Tree Frog	X	X				X	X	X	X	X	X
(Litoria adelaidensis)											
Green and Gold Tree Frog	X	X	X			X		X	X		X
(Litoria moorei)											
Family Myobatrachidae											
Western Banjo Frog	X		X	X			X	X	X	X	X
(Limnodynastes dorsalis)											
Burrowing Frog		X		X		X	X		X		X
(Heleioporus inornatus											
Moaning Frog	X	X	X	X	X	X	X	X	X		X
(Heleioporus eyeri)											
Heleioporus psammophilus	X						X				
Humming Frog						X					
(Neobatrachus pelobatoides)											
Pseudophryne nichollsi				X		X		X		X	X
Gunther's Toadlet	X			X	X				X		
(Pseudophrne guntheri)											
Crinia georgiana	X	X		X	X	X		X	X	X	X
Ranidella glauerti		X		X		X	X	X		X	X
Ranidella insignifera						X					
Ranidella subinsignifera	X							X			
Ranidella pseudinsignifera											
Geocrinia leai					X	X	X	X	X		
Geocrinia alba											X
Geocrinia lutea										X	
Geocrinia rosea							X	X			
Geocrinia vitellina											X

NOTE: All species of vertebrates listed are recent records having been recorded in the area during the last 20 years.
[a] Other - includes species recorded outside the forested areas in farmland and pine plantations.

Appendix 3e FRESHWATER FISH RECORDED IN MAJOR NATIONAL PARKS AND NATURE RESERVES

LEGEND

1. D'Entrecasteaux (NP)
2. Perup (NR)
3. Leeuwin-Naturaliste (NP)
4. St Johns Brook (CP)-
   Whicher Range (NR)
5. Milyeannup (NR)
6. Mt. Frankland (NP)
7. Shannon (NP)
8. Warren (NP)
9. Mt. Lindesay (NP)
10. Walpole-Nornalup (NP)
11. Other[a]

NP = National Parks
NR = Nature Reserves
CP = Conservation Parks

	NATIONAL PARKS OR RESERVES										
	1	2	3	4	5	6	7	8	9	10	11
Family Galaxiidae											
Western Minnow (*Galaxias occidentalis*)	X			X		X	X	X	X	X	X
Black-Striped Minnow (*Galaxiella nigrostriatus*)							X			X	X
Mud Minnow (*Galaxiella munda*)						X					
Family Lepidogalaxiidae											
Salamander Fish (*Lepidogalaxias salamandroides*)	X					X	X			X	X
Family Percichthyidae											
Night Fish (*Bostockia porosa*)	X			X	X	X	X	X	X	X	X
Family Kuhliidae											
Western Pygmy Perch (*Edelia vittata*)	X			X	X	X	X	X			X
Balstons Pygmy Perch (*Nannatherina balstoni*)	X						X	X			X
Family Plotsidae											
Freshwater Cobbler (*Tandanus bostocki*)	X							X			X
Family Geotriidae											
Pouched Lamprey (*Geotria australis*)								X			X
Family Atherinidae											
Hardyhead (*Atherinosoma edelensis*)											X

NOTE: All species of vertebrates listed are recent records having been recorded in the area during the last 20 years.
[a] Other - includes species recorded outside the forested areas in farmland and pine plantations.

	1	2	3	4	5	6	7	8	9	10	11
Family Mugilidae											
Mangrove Mullet *(Mugil cephalus)*	X						X				
Family Gobiidae											
Swan River Goby *(Pseudogobius olorum)*									X		
EXOTIC FISH											
Family Poecilidae											
Mosquito Fish *(Gambusia affinis)*						X					X
Family Cyprinidae											
Carp *(Carassius carassius)*											X
Family Percidae											
Redfin Perch *(Perca fluviatilis)*						X					X
Family Salmonidae											
Brown Trout *(Salmo trutta)*	X							X			
Rainbow Trout *(Salmo gairdneri)*								X			

Appendix 3f INSECTS RECORDED IN THE SOUTHERN FORESTS
(extracted from Abbott in press)[a]

COLLEMBOLA

PODURIDAE
Achorutes rosaceus Schott
* *Hypogastrura armata* Nicolet
Hypogastrura pseudopurpurascens Womersley

ISOTOMIDAE
Acanthocyrtus plumbeus (Womersley)
* *Cryptopygus thermophilus* (Axelson)
** *Isotomurus chiltoni* (Carpenter)

ENTOMOBRYIDAE
Entomobrya varia Schött
* *Sinella coeca* (Schött)

SMINTHURIDAE
Arrhopalites coecus (Tullberg)
Deuterosminthurus handschini Womersley
Megalothorax swani Womersley
Parakatianna pulchella Womersley
Sminthurides stagnalis Womersley

PROTURA

EOSENTOMIDAE
Eosentomon swani Womersley

DIPLURA

CAMPODEIDAE
Metriocampa spinigera Womersley

JAPYGIDAE
Japyx froggatti Silvestri
Japyx nichollsi Womersley
Japyx tillyardi Silvestri

EPHEMEROPTERA

LEPTOPHLEBIIDAE
Neboissophlebia occidentalis Dean

ODONATA

COENAGRIONIDAE
Austroagrion cyane (Selys)
Ischnura aurora (Brauer)

LESTIDAE
Austrolestes annulosus (Selys)

MEGAPODAGRIONIDAE
Argiolestes minimus Tillyard

AESHNIDAE
Aeshna brevistyla (Rambur)
Austroaeschna anacantha Tillyard
Hemianax papuenais (Burmeister)

GOMPHIDAE
Armagomphus armiger (Tillyard)
Austrogomphus collaris Hagen
Austrogomphus lateralis (Selys)

CORDULIIDAE
Hesperocordulia berthoudi Tillyard
Lathrocordulia metallica Tillyard
Procordulia affinis (Selys)
Synthemis leachii Selys
Synthemis macrostigma Hagen

LIBELLULIDAE
Austrothemis nigrescens (Martin)
Diplacodes bipunctata (Brauer)
Nannophya dalei Tillyard
Nannophya occidentalis (Tillyard)

BLATTODEA

BLATTIDAE
Allacta similis (Sauss.)
Drymaplaneta heydeniana (Saussure)
Drymaplaneta semivitta (Walker)
Eppertia punctata Princis
Melanozosteria morosa Shelford
Platyzosteria fallax (Walker)
Platyzosteria fulva Princis
Platyzosteria obscura (Tepper)
Polyzosteria cuprea Saussure

BLATELLIDAE
Ceratinoptera ensifera Lunenberg,
Hensaussurea halmaturina (Tepper)
Hensaussurea sheardi Princis
Johnrehnia hodgkini Princis
Neotemnopteryx fulva (Saussure)
Robshelfordia hartmani Roth

ISOPTERA

KALOTERMITIDAE
Kalotermes hilli Emerson

RHINOTERMITIDAE
Coptotermes acinaciformis (Froggatt)
Cryptotermes austrinus Gay & Watson
Heterotermes platycephalus Froggatt

---

* Introduced species

** Species recorded from Porongurup National Park but not from the rest of the southern forests

[a]ABBOTT, I. (in press). Prodromus of the occurrence and distribution of insect species in the forested part of south-west Western Australia. Department of Conservation and Land Management (W.A.) Technical Report.

TERMITIDAE
Ahamitermes hillii (Nicholls)
Amitermes obeuntis (Silvestri)
Occasitermes occasus (Silvestri)
Paracapritermes hesperus Gay
Peracapritermes kraepelinii (Silvestri)

DERMAPTERA

PYGIDICRANIDAE
Dacnodes shortridgei (Burr)

ANISOLABIDIDAE
Anisolabis westralica Burr
Gnolabis woodwardi Burr

PLECOPTERA

GRIPOPTERYGIDAE
Leptoperla australica (Enderlein)
Newmanaperla thoreyi (Banks)

ORTHOPTERA

TETTIGONIIDAE
Mygalopsis pauperculus (Walker)
Mygalopsis sandowi Bailey
Tympanophora pellucida White

GRYLLIDAE
Balamara albovittata (Chopard)
Bobilla bivittata (Walker)
Endacusta pindana Otte & Alexander
Nambungia balyarta Otte & Alexander

PYRGOMORPHIDAE
Monistria discrepans (Walker)
Monistria maculicornis Sjöstedt
Psedna nana (Rehn)

ACRIDIDAE
Chortoicetes terminifera (Walker)
Gastrimargus musicus (Fabricius)
Phaulacridium vittatum (Sjöstedt)

TETRIGIDAE
Cyphotettix camelus Rehn
Paratettix australis (Walker)
Peronotettix cyclopyga Rehn

PSOCOPTERA

PHILOTARSIDAE
Austropsocus occidentalis Thornton & New
Haplophallus guttatus (Tillyard)

MYOPSOCIDAE
Phlotodes australis (Brauer)

HEMIPTERA

PSYLLIDAE
Anoeconeossa vespertina Taylor
Cryptoneossa occidentalis Taylor
Glycaspis icterica Moore

ALEYRODIDAE
Xenaleyrodes eucalypti Dumbleton

APHIDIDAE
Anomalaphis comperei Pergande
Ceriferella dossuaria Carver & Martyn

ADELGIDAE
* *Pineus pini* (Macquart)

PSEUDOCOCCIDAE
Eucalyptococcus gisleni (Ossiannilsson)
Maconellicoccus lanigerus (Fuller)

ERIOCOCCIDAE
Apiomorpha maliformis Fuller
Apiomorpha subconica (Tepper)

DIASPIDIDAE
Incertae sedis
Bunduica rubrovenosa Jacobi

CICADELLIDAE
Rosopaella nigropicta Webb

EURYMELIDAE
Eurymeloides punctata (Signoret)
Eurymelops rubrovittata (Amyot & Serville)

MEMBRACIDAE
Sextius depressus Goding

OCHTERIDAE
Megochterus occidentalis Baehr

GELASTOCORIDAE
Nerthra femoralis (Montandon)
Nerthra stali (Montandon)
Nerthra tuberculata (Montandon)

MIRIDAE
Chaetedus longiceps Eyles

REDUVIIDAE
Ptilocnemus lemur (Westwood)

ARADIDAE
Aspisocoris termitophilus Kormilev
Neuroctenus majusculus Bergroth

LYGAEIDAE
Austronysius sericus Ashlock
Austroxetes westraliensis Woodward
Botocudo aff. ornatulus (Bergroth)

Crompus opacus Scudder
Cryptorhamphus orbus Stal,
Dieuches notatus (Dallas)
Euander lacertosus (Erichson)
Euander multicoloratus (Distant)
Ischnodemus sordidus Slater
Isopeltus australis (Bergroth)
Lepionysius grossi Ashlock
Nysius vinitor Bergroth
Porander scudderi Gross
Spilostethus pacificus (Boisduval)
Udeocoris nigroaeneus (Erichson)

COREIDAE
Amorbus bispinus (Westwood)

SCUTELLERIDAE
Choerocoris paganus (Fabricius)

PENTATOMIDAE
Chaetoceris variegatus
Cuspicona procallosa Gross
Nezara viridula (Linnaeus)
Omyta spinosa (Van Duzee)
Poecilometis lineatus (Westwood)
Poecilometis punctiventris (Stal)
Pseudapines geminata (Van Duzee)

THYSANOPTERA

AEOLOTHRIPIDAE
Desmothrips steeleae Mound
Haplothrips victoriensis (Bagnall)
Phaulothrips fuscus (Moulton)

THRIPIDAE
Chirothrips manicatus Haliday
Odontothripiella passalaina Pitkin
Scolothrips sexmaculatus (Pergande)
Taeniothrips breviocornis (Bagnall)
Thrips australis Bagnall
Thrips imaginis (Bagnall)
Thrips tabaci (Lind.)

PHLAEOTHRIPIDAE
Apterygothrips australis Pitkin
Baenothrips moundi (Stannard)
Carientothrips badius (Hood)
Carientothrips vesper Mound
Haplothrips angustatus Hood
Nesothrips propinquus (Bagnall)

MEGALOPTERA

CORYDALIDAE
Archichauliodes cervulus Theischinger

NEUROPTERA

ITHONIDAE
Ithone westraliensis Riek

MANTISPIDAE
Campion tenuistriga (Gerstaecker)
Theristria basalis Banks

HEMEROBIIDAE
Zachobiella lobata New
Myrmeleon acer Walker
Periclystus aureolatus Tillyard

CHRYSOPIDAE
Calochrysa extranea (Esben-Petersen)
Chrysopa edwardsi Banks
Chrysopa ramburi Schneider
Chrysopa signata Schneider
Chrysopa tripunctata McLachlan
Dictyochrysa peterseni Kimmins
Nothancyla verreauxi Navás

ASCALAPHIDAE
Acmonotus incusifer McLachlan
Megacmonotus magnus (McLachlan)

CARABIDAE

CARABINAE
Agonocheila fenestrata Blackburn
Cerotalis substriata Castelnau
Coptocarpus thoracicus (Castelnau)
Demetrida constricticeps (Sloane)
Harpalus fulvicornis (Thunberg)
Mecyclothorax ovalis Sloane
Microferonia adelaidae Blackburn
Microferonia cinctipennis Sloane
Neonomius australis (Sloane)
Notonomus mediosulcatus (Chaudọir)
Pentagonica vittipennis Chaudoir
Pericompsus habitans (Sloane)
Promecoderus albaniensis Castelnau
Pseudoceneus sollicitus (Erichson)
Sarticus elegantulus (Castelnau)
Scopodes boops Erichson
Simodontus australis (Dejean)
Simodontus occultus Sloane
Simodontus sexfoveatus Chaudoir)
Techimorphus westraliensis Moore
Trechobembix baldiensis (Blackburn)
Trigonothops pescotti (Oke)

DYTISCIDAE
Allomatus nannup Watts
Copelatus ater Sharp
Liodessus dispar (Sharp)
Liodessus inornatus (Sharp)
Necterosoma darwini (Babington)
Platynectes aenescens Sharp
Sternopriscus marginatus Watts

GRYRINIDAE
Aulonogyrus strigosus (Fabricius)
Macrogyrus angustatus Régimbart

HYDROPHILIDAE
Berosus approximans Fairmaire
Chledocyon intermedius Hansen
Chledocyon marmoratus Hansen
Chledocyon semiopacus Hansen
Chledocyon venustus Hansen

HISTERIDAE
Acritus blighi Gomy
Acritus occidentalis Lea
Acritus tropicus Lea
Bacanius suturalis (Lea)
Platysoma strangulatum Mars.

LEIODIDAE
Austronargus semota (Szymczakowski)
Myrmecholeva lata Lea
Myrmecholeva punctata Lea

SCYDMAENIDAE
Phagonophana latipennis Lea

STAPHYLINIDAE
Aleochara insignis Blackburn
Catodera alternans Lea
Calodera marginicollis Lea
Conosoma barycephalum Lea
Conosoma limnorioides Lea
Conosoma orthodoxum Lea
Hetairotermes punctiventris (Lea)
Homalota australasiae Lea
Hyperomma nigrum Lea
Leptacinus bisulciceps Lea
Oedichirus andersoni Blackburn
Oxytelus cribriceps Fauv.
Paederus meyricki Blackburn
Scaphisoma instabile Lea
Stenus guttulifer Waterhouse
Stenus bifenestratus L. Benick
Termophila punctiventris Lea

PSELAPHIDAE
Articerus cultripes Raffray
Palimbolus dimidiatus (Raffray)
Euplectops discopallidus Lea
Eupines flavoapicalis Lea
Plectostenus gracilicornis Lea
Pselaphus pilosus Raffray
Pselaphus tuberculifrons Raffray
Tmesiphorus ponerae Lea

CLAMBIDAE
Clambus simsoni Blackburn

TROGIDAE
Trox alternans Macleay
Trox costatus Wiedemann

GEOTRUPIDAE
Blackbolbus taurus (Westwood)
Bolborhacium bainbridgei (Westwood)
Bolborhacium hollowayi Howden
Bolborhacium recticorne (Guérin)
Bolborhacium triunum (Lea)
Stenaspidius nigricornis Westwood

SCARABAEIDAE

APHODIINAE

Aphodius frenchi Blackburn
Aphodius granarius L.
* *Aphodius pseudolividus* Balthazar

SCARABAEINAE
***Labroma horrens* Sharp
Labroma umbratilis Matthews
Lepanus occidentalis Matthews
Onthophagus evanidus Frey
Onthophagus ferox Harold
Onthophagus haagi Harold
Onthophagus rupricapra Waterhouse
Onthophagus vermiculatus Frey
Proctophanes sculptus Hope
Sauvagesinella backi (Paulian)

MELOLONTHINAE
Ateromonocheila longipes
Automolus immitis (Blackburn)
Biphyllocera kirbyana White
Byrrhomorpha ponderosa Blackburn
Colpochila amabilis (Blackburn)
Colpochila efficax Britton
Colpochila nobilis (Blackburn)
Colpochilodes peregrina Britton
Colpochilodes raucipennis Blackburn
Colymbomorpha lineata Blanchard
Diphucephala edwardsi Waterhouse
Enamillus rectus Allsopp
Haplopsis polita Britton
Heteronyx agrestus Burmeister
Heteronyx nudus Blackburn
Heteronyx nigricans Burmeister
Liparetrus collessi Britton
Liparetrus gravidus Blackburn
Liparetrus jenkinsi Britton
***Liparetrus mulurus* Britton
Maechidius major Blackburn
Neophyllotocus legnotus Britton
Phyllococerus purpurascens Waterhouse
Phyllotocus ustulatus Blanchard
Prochelyna heterodoxa Burmeister
Sericesthis fovea Britton
Sphaeroscelis pectoralis Burmeister
Synchilus gisleni Britton

RUTELINAE
Anoplostethus opalinus Brullé
Epichrysus lamprimoides White

DYNASTINAE
Cryptodus costulipennis Fairmaire
Cryptodus goerlingi Carne

Cryptodus passaloides Germar
Cryptodus variolosus White
Heteronychus arator Fabricius
Novapus crassus Sharp
Semanopterus angustatus Blackburn
Semanopterus leai Blackburn
Trissodon curtus (Burmeister)

CETONIINAE
Micropoecila breweri (Janson)
Trichaulax philipsii (Schreibers)

BUPRESTINAE
***Castiarina amabilis* (Gory & Laporte)
Castiarina decemguttata (Gory)
Catiarina discolorata (Barker)
Castiarina flavopicta (Boisduval)
Castiarina rufipennis (Kirby)
Melobasis cuprifera Gory & Laporte
Melobasis wannerua Carter
Diadoxus erythrurus (White)

POLYCESTINAE
***Astraeus flavopictus* LaPorte & Gory

ELATERIDAE
Crepidomenus dysmikos Calder
Crepidomenus occidualis Calder
Lacon caliginosus Guer.
Lingana dombarta Neboiss
Paracardiophorus occidentalis Carter

CANTHARIDAE
Heteromastix flavipennis Lea
Heteromastix pusillus Boh.

DERMESTIDAE
Dryophilodes latipennis Lea
Hadrobregmus australiansis Pic

BOSTRICHIDAE
Xylopsocus rubidus Lesne

LYMEXYLIDAE
Atractocerus crassicornis Clark
Atractocerus kreuslerae Pascoe

TROGOSSITIDAE
Ancyrona lewisi Reitt.

CLERIDAE
Opilo congruus Newman
Trogodendron fasciculatum Schreibers

MELYRIDAE
Carphurus cyanopterus Boh.
Helcogaster parallelus Lea
Dicranolaius intermedius Lea

PHALACRIDAE
Litochrus sydneyensis Blackburn
Parasemus australiae Lea
Phalacrus halacrinus Blackburn

BIPHYLLIDAE
Diplocoelus latus Lea

BOTHRIDERIDAE
Deretaphrus piceus Germ.

CERYLONIDAE
Euxestus medioniger Lea

COCCINELLIDAE
Cleobora mellyi (Mulsant)
Coccinella transversalis Fabricius
Harmonia conformis (Boisduval)
Rhyzobiellus alphabeticus Lea
Stethorus vagans Blackburn

TENEBRIONIDAE
Adelium forticorne Gebien
Adelium irregulare Carter
Adelium vicarium Pascoe
Anaxo cylindricollis Carter
Dimorphochilus apicalis Borchmann
Dimorphochilus gouldi Hope
Ecnolagria aeneoviolacea Champ.
Helaeus perforatus
Helaeus rugosipennis Carter
Melaps pilosus Carter
Omolipus grimwadei Oke
Ophrythyreocis vigilans Lea
Platydema abdominale Gebien

MORDELLIDAE
Mordella brevis Lea

RHIPIPHORIDAE
Evaniocera meyricki Blackburn

ZOPHERIDAE
Latometus lunatus Pascoe

OEDEMERIDAE
Copidita puncta W.S. Macleay

ANTHICIDAE
Anthicus geminatus Lea
Anthicus simulator Lea

ADERIDAE
Syzetoninus parallelus Lea

CERAMBYCIDAE

PRIONINAE
Scelecantha pilosicollis (Hope)

CERAMBYCINAE
Adrium artifex (Newman)

Amphirhoe decora (Newman)
Aphanasium australe (Boisduval)
Bardistus cibarius Newman
Bethelium cleroides (White)
Bethelium ruidum Pascoe
Bimia bicolor White
Coptoceraus rubripes (Boisduval)
Coptocercus undulata (Hope)
Homaemota basalis (Pascoe)
Ochyra variabilis Lea
Phoracantha impavida Newman
Phoracantha lata (Hope)
Phoracantha semipunctata (Fabricius)
Pseudocephalus mirus (Pascoe)
Stenocentrus suturalis (Olivier)
Syllitus grammicus (Newman)
Tryphocaria acanthocera (Macleay)
Tryphocaria punctipennis Blackburn
Uracanthus triangularis Hope

LAMIINAE
Ancita didyma Blackburn
Ancita marginicollis (Boisduval)
Ancita sparsa (Pascoe)
Disternopsis pentheodes (Pascoe)

CHRYSOMELIDAE

CHRYSOMELINAE
Calomela maculicollis Boisduval
Chalcolampra arthritica Lea
Chalcolampra consimilis Lea
Chrysophtharta amoena Chapuis
Chrysophtharta decolorata (Chapuis)
Cyclonoda pilula Clark
Paropsis advena
Paropsis elliptica
Paropsis elytrura Blackburn
Paropsis festiva Chap.
Paropsis mentitrex Blackburn
Paropsis yilgarnensis Blackburn
Strumatophyma verrucosa Clark

CRYPOTCEPHALINAE
Cadmus nothus Lea
Ditropidus brunneipennis Lea
Ditropidus cornutus Baly
Ditropidus flavoapicalis Lea
Ditropidus fugitivus Chap.
Ditropidus impuncticollis Lea
Ditropidus pulicosus Lea
Ditropidus xanthurus Lea
Elaphodes halticoides Lea
Leasia australis Jacoby
Loxopleurus lateriflavus Lea
Loxopleurus mixtus Lea

SAGRINAE
Diaphanops westermanni (Boheman)

EUMOLPINAE
Agetinella minuta Jacoby
Edusella blackburni Lea
Edusella dispar Lea
Geloptera geniculata Baly
Geloptera soror Lea
Geloptera tuberculata Baly
Tomyris picticornis Lea
Tomyris pusilla Lef.

GALERUCINAE
Arsipoda acuminata Waterhouse
Monolepta hypomela Lea

CURCULIONIDAE

AMYCTERINAE
***Acantholophus amycteroides* Macleay
Acantholophus aureolus Boheman
Acantholophus bivittatus Boheman
Acantholophus dumosus Boheman
Acantholophus hypoleucus Boheman
Acantholophus lateralis Boheman
Acantholophus maximus Macleay
Acantholophus mucronatus Macleay
Acantholophus rugiceps Macleay
Acantholophus scaphirostris Ferguson
Acantholophus spinosus Macleay
***Acantholophus suturalis* Boheman
Aedriodes crawshawi Ferguson
Aedriodes humeralis Lea
Aedriodes inuus Pascoe
Aedriodes mendosus Pascoe
Aedriodes mucronatus Ferguson
Cucullothorax horridus Ferguson
Dialeptopus collaris Boheman
Dialeptopus ferreus Pascoe
Dicherotropis dameli Macleay
Euomus insculptus Boheman
Euomus scorpio Boisduval
Euomus stephensi Gyll.
Hyborrhynchus coenosus
Hyborrhynchus maculatus Macleay
Macramycterus boisduvali Boisduval
Melanegis stygius Pascoe
Mythitis basalis Boisduval
Neohyborrhynchus rugosus Macleay
Oditesus sulcirostris Pascoe
Parahyborrhynchus convexiusculus Macleay
Sclerorinus carteri Ferguson
Talaurinus angustatus Macleay
Talaurinus semispinosus Boheman

LEPTOPIINAE
Catasaurcus aerosus Thompson
Catasarcus bilineatus Fahraeus
Catasarcus coruscus Thompson
Catasarcus cygnensis Thompson
Catasarcus frontalis Thompson
Catasarcus hopei Fahraeus
Catasarcus impressipennis (Boisduval)
Catasarcus laevior Thompson
Catasarcus murex Thompson
Catasarcus spinipennis Fahraeus

Cherrus mastersi Pascoe
Essolithna echimys Pascoe
Leptopius spinosus Lea
Polyphrades aesalon Pascoe

GONIPTERINAE
Oxyops farinosa Pascoe
Oxyops fasciata Boisduval
Oxyops pictipennis Blackburn

DIABATHRARIINAE
Strongylorhinus clarki Marshall

RHYNCHITINAE
Auletes pallipes Lea

RHADINOSOMINAE
Rhadinosomus lacordairei Pascoe

ERIRHININAE
Cydmaea eucalypti Lea
Cydmaea uniformis Lea
Cyttalia maculata Lea
Cyttalia sydneyensis Blackburn
Desiantha trivitticollis Lea
Dicrocis leucomelas Lea
Omorophius nigrovarius Lea
Storeus macrostylus Lea

LAEMOSACCINAE
Laemosaccus cryptonyx Pascoe

OTIORHYNCHINAE
**Otiorrynchus cribricollis* Gyll.
Pascoellus umbratus Blackburn

AMALACTINAE
Tranes roei Boheman
Tranes xanthorrhoeae Lea

CRYPTORHYNCHINAE
Achopera lachrymosa Pascoe
Decilaus distans Pascoe
Decilaus foraminosus (Pascoe)
Decilaus moluris Lea
Lybaeba oestuans Pascoe
Menios nebulosus Lea
Ophrythyreocis vigilans Lea
Tetengia solenopa Lea

RHYNCHAENINAE
Rhamphus microscopicus Lea

HAPLONYCHINAE
Haplonyx tibialis

TYCHIINAE
Elleschodes inconstans Lea
Elleschodes placidus Lea

STREPSIPTERA

STYLOPIDAE
Austrostylops gracilipes Lea

MECOPTERA

BITTACIDAE
Harpobittacus phaeoscius Riek
Harpobittacus similis Esben-Petersen

MEROPEIDAE
Austromerope poultoni Killington

DIPTERA

TIPULIDAE
Amphineurus fergusoni Alexander
Limonia punctipennis (Skuse)
Helius invenustipes (Alexander)
Horistomyia occidentalis Alexander
***Leptotarsus mathewsi* (Alexander)
Leptotarsus walpoleiensis Dobrotworsky
Paralimnophila indecora (Alexander)
Limonia clarki Alexander
Molophilus inimicus Alexander
Ptilogyna lineata (Skuse)
Ptilogyna westralis Dobrotworsky

TANYDERIDAE
Radinoderus occidentalis (Alexander)

CULICIDAE
Aedes alboannulatus (Macquart)
Aedes camptorhynchus (Thomson)
Aedes clelandi (Taylor)
Aedes notoscriptus (Skuse)
Aedes occidentalis (Skuse)
Aedes ratcliffei Marks
Anopheles annulipes Walker
Anopheles atratipes Skuse
Culex annulirostris Skuse
Culex fatigans Wiedemann
Culex aff. fergusoni (Taylor)
Culex pipiens Linnaeus
Culiseta atra (Lee)

CHIRONOMIDAE
Ablabesmyia notabilis (Skuse)
Cladotanytarsus australomancus Glover
Kiefferulus intertinctus (Skuse)
***Kiefferulus martini* (Freeman)
Paramerina levidensis (Skuse)
Polypedilum nubiferum (Skuse)
Procladius paludicola Skuse
Tanytarsus collessi Glover
Tanytarsus edwardi Glover
Tanytarsus inextentus Skuse
Tanytarsus fuscithorax Skuse

CERATOPOGONIDAE
Leptoconops longicornis Carter
Mixohelea lemur Debenham
Mixohela versicolor Debenham

SIMULIIDAE
Austrosimulium bancrofti (Taylor)
Austrosimulium aff. cornutum Tonnoir
Austrosimulium furiosum (Skuse)
Cnephia(?) fergusoni (Tonnoir)
Cnephia(?) aff. terebrans (Tonnoir)
Simulium ornatipes Skuse

PSYCHODIDAE
Brunettia subalternata (Satchell)
Psychoda squamipleuris Satchell
Peripsychoda crassepalpis (Satchell)
Telmatoscopus norrisi Satchell

BIBIONIDAE
Dilophus cantrelli Hardy
Dilophus collessi Hardy
Dilophus tricuspidatus Hardy

PELECORHYNCHIDAE
Pelecorhynchus mackerrasi Daniels

RHAGIONIDAE
Atherimorpha commoni Paramonov
Dasyomma norrisi Paramonov
Spania rieki Paramonov

TABANIDAE
Mesomyia norrisi Mackerras
Scaptia bicolorata (Taylor)
Scaptia calabyi Mackerras
Scaptia gemina (Walker)
Scaptia georgii (Taylor)
Scaptia gibbula (Walker)
Scaptia neotricolor (Taylor)
Scaptia singularis (Macquart)
Scapta tricolor (Walker)

NEMESTRINIDAE
Trichophthalma fortei Paramonov

ASILIDAE
Amphisbetetus primus Paramonov
Hullia commoni Paramonov
Mauropteron pelago (Walker)

BOMBYLIIDAE
Alencosia affinis Yeates
Alencosia arguta Yeates
Aleucosia cincta (Edwards)
Aleucosia corculum (Newman)
Aleucosia dorsalis (Walker)
Aleucosia maculosa (Newman)
Aleucosia plena (Walker)
Aleucosia saltuaria Yeates
Aleucosia tridentata Yeates

Oncodasia patula (Walker)

SYRPHIDAE
Microdon occidentalis Ferguson

LAUXANIIDAE
Poecilohetaerus albolineatus Hendel
Poecilohetaerus aquilus Schneider

PLATYSTOMATIDAE
Duomyia capnodes McAlpine
Duomyia dete McAlpine
Duomyia lonchaeina McAlpine
Duomyia longicauda McAlpine
Duomyia lutea McAlpine
Duomyia ustulata McAlpine
Lenophila nila McAlpine & Kim

MICROPEZIDAE
Badisis ambulans McAlpine

HELEOMYZIDAE
Diplogeomyza maculipennis (Malloch)
Tapeigaster fulva Malloch
Tapeigaster nigricornis (Macquart)
Tapeigaster paramonovi McAlpine & Kent

AGROMYZIDAE
***Butomomyza poemyzina* (Spencer)
Cerodontha milleri Spencer
Fergusonina lockharti Tonnoir
Melanagromyza seneciophila Spencer
Melanagromyza variegata Spencer
Ophiomyia fera Spencer
Phytoliriomyza nigricans Spencer
Phytomyza anthocecidis Spencer
Phytomyza clematidicolla Spencer
Phytomyza ranunculicaulis Spencer
Phytomyza vitalbae Kaltenbach

EPHYDRIDAE
Ephydrella acrostichalis (Malloch)
Hydrellia perplexa Bock
Hydrellia tritici Coquillet
Hydrellia victoria Cresson

DROSOPHOLIDAE
Drosophila fuscithorax Malloch
Drosophila grossfieldi Bock
Drosophila nicholsoni Malloch
Scaptomyza australis Malloch
Scaptomyza pallida (Zetterstedt)

CHLOROPIDAE
Chloromerus maculifemur Malloch
***Chloromerus purus* Becker
Diplotoxa sexta Spencer
Tricimba selachopina Thomson

MUSCIDAE
Atherigona tibiseta Malloch
Australophyra rostrata Robineau-Desvoidy

Fannia australis Malloch
Gymnodia obliterata (Malloch)
Musca vetustissima Walker
Prohardyia carinata (Stein)
Prohardyia pollinosa (Malloch)

CALLIPHORIDAE
Calliphora varifrons Malloch

TACHINIDAE
Formosiomima nigromaculata (Malloch)
***Chaetophthalmus dorsalis* (Malloch)
Chrysopasta elegans (Macquart)
Cylindromyia bimacula (Walker)
***Cylindromyia brunnea* Malloch
Prodiaphania cygnus (Malloch)
Quadra ornata Malloch
Winthemia lateralis (Macquart)

TRICHOPTERA

HYDROBIOSIDAE
Apsilochorema urdalum Neboiss
Taschorema pallescens (Banks)

HYDROPTILIDAE
Hellyethira litua Wells
Hydroptila losida Mosely
Oxyethira brevis Wells
Oxyethira retracta Wells

PHILOPOTAMIDAE
Hydrobiosella amblyopia Neboiss
Hydrobiosella michaelseni (Ulmer)

HYDROPSYCHIDAE
Cheumatopsyche modica (McLachlan)
Smicrophylax australis (Ulmer)

POLYCENTROPODIDAE
Adectophylax volutus Neboiss
Plectrocnemia eximia Neboiss

ECNOMIDAE
Ecnomina merga Neboiss
Ecnomina mesembria Neboiss
Economina scindens Neboiss
Ecnomina sentosa Neboiss
Ecnomina viatica Neboiss
Ecnomus pansus Neboiss
Ecnomus turgidus Neboiss

PLECTROTARSIDAE
Plectrotarsus minor Mosely

PHILORHEITHRIDAE
Kosrheithrus boorarus Neboiss

LEPTOCERIDAE
Condocerus aptus Neboiss
Lectrides parilis Neboiss
Notalina delicatula Ulmer
Notalina flava (Ulmer)
Notalina fulva Kimmins
Notalina pseudodelicata Neboiss
Notoperata tenax Neboiss
Oecetis cymula Neboiss
Oecetis laustra Mosely
Oecetis pechana Mosely
Oecetis walpolica Neboiss
Symphitoneuria wheeleri Banks
Triplectides australis Navás
Triplectides enthesis Neboiss

LEPIDOPTERA

HEPIALIDAE
Abantiades hydrographus (Felder)
Abantiades ocellatus (Tindale)
Aenetus dulcis (Swinhoe)
Aenetus scriptus (Scott)
Fraus distispina Nielsen & Kristensen
Fraus furcata Nielsen & Kristensen
Fraus mediaspina Nielsen & Kristensen
Fraus pilosa Nielsen & Kristensen
Fraus pteromela (Lower)
Fraus serrata Nielsen & Kristensen
Oxycanus kochi Tindale
Oxycanus occidentalis Tindale
Oxycanus poeticus Tindale
Oxycanus promiscuus Tindale

INCURVARIIDAE
Perthida glyphopa Common

PSYCHIDAE
Lomera pantosemna (Turner)
Lomera xanthochrysa (Meyrick & Lower)

TINEIDAE
Scieropepla monoides Turner

OECOPHORIDAE
Baryzancla dysclyta Turner
Blacophanes pallida Turner
Myrascia interlineata (Walker)
Myrascia subductella (Walker)
Philetes megalospila Turner
Thudaca campylota Meyrick

COSSIDAE
Archaeoses polygrapha Lower
Xyleutes cinereus Turner
Xyleutes vittata Walker

TORTRICIDAE
Arotrophora anemarcha (Lower)
Arotrophora arcuatalis (Walker)
Arotrophora diadela Common
Palaeotoma styphelana Meyrick
Paraphyas callixena Turner
Parastranga macrogona Meyrick
Tortrix divulsana Walker

LIMACODIDAE
Doratifera oxleyi (Newman)
Doratifera quadriguttata Walker
Doratifera vulnerans (Lewin)

PYRALIDAE
Hednota icelomorpha Turner
Hednota pedionoma (Meyrick)
Hednota recurvella (Walker)

GEOMETRIDAE
Acidalia rubraria Dbld
Adeixis inostentata (Walker)
Boarmia leucanthes Turner
Chlenias basichorda Turner
Chlenias macrochorda Turner
Chlorocoma rhodocrossa (Turner)
Dichromodes obtusata (Walker)
Euchloris rhodocrossa Turner
Idiodes apicata Guenée
Mnesampela arida McQuillan
Mnesampela privata (Guenée)
Nearcha staurotis Meyrick
Nycterephes coracopa Turner
Poecilasthena oceanias Meyrick
Prasinocyma semicrocea Walker
Taxeotis thegalea Turner
Thalaina angulosa Walker

LYCAENIDAE
Candalides hyacinthinus (Semper)

LASIOCAMPIDAE
Digglesia rufescens Walker

ANTHELIDAE
Munychrygia periclyta Common & McFarland
Pterolocera amplicornis Walker

SATURNIDAE
Antheraea helena (White)

NOTODONTIDAE
Discophlebia lipauges Turner
Ochrogaster lunifer Herrich-Schaffer
Oenosandra boisduvalii Newman

LYMANTRIIDAE
Acyphas semiochrea (H.-Sch.)
Teia athlophora (Turner)

NOCTUIDAE
Heliothis punctigera Wallengren
Spathoptila cyclophora Turner
Thoracolopha pissonephra Turner
Uraba lugens Walker
Zia tactalis Walker

HYMENOPTERA

PERGIDAE
Antiperga antiopa Morice
Lophyrotoma analis (Costa)
Perga brullei Westwood
Perga christii (Westwood)
Perga klugii Westwood
Perga konowi (Benson)
Perga mayri Westwood
Perga schiodtei Westwood
**Phylacteophaga froggatti* Riek

MEGALYRIDAE
Megalyra fusciipennis Westwood

AULACIDAE
Aulacostethus cingulatus (Westwood)

ICHNEUMONIDAE
Enicospilus amplipennis (Morley)
Enicospilus gardei (Morley)
Enicospilus pseudantennatus Gauld
Enicospilus pulkus Gauld
Leptophion antennatus (Morley)

BRACONIDAE
Syngaster rubricepsis (Shenefelt)
Miropotes thuraris Austin

DIAPRIIDAE
Diphoropria collessi Naumann
Diphoropria nigricans (Dodd)
Diphoropria tricarinata Naumann

MONOMACHIDAE
Monomachus hesperius Naumann

PLATYGASTRIDAE
Parabaeus abyssus Austin

PTEROMALIDAE
Anepistenia vexans Boucek
Enoggera nassaui (Girault)
Enoggera reticulata Naumann
Enoggera sulcata Naumann

POMPILIDAE
Psoropempula erthrostethus (Smith)
Turneromyia melancholica (Smith)
Turneromyia vassei (Turner)
Turneromyia venator Evans

SCOLIIDAE
Anthobosca clypeata Smith

FORMICIDAE

MYRMECIINAE
Myrmecia analis Mayr
Myrmecia chasei Forel
Myrmecia clarki Crawley
Myrmecia infima Forel
Myrmecia mandibularis Smith
Myrmecia michaelseni Forel

Myrmecia nigra Forel
Myrmecia nigriceps Mayr
Myrmecia regularis Crawley
Myrmecia rugosa Wheeler
****Orectognathus clarki* Brown

PONERINAE
Amblyopone australis Erichson
Discothyrea crassicornis Clark
Rhytidoponera metallica (Smith)
Rhytidoponera punctigera Crawley

MYRMICINAE
Stromigenys quinquedentata Crawley

DOLICHODERINAE
Iridomyrmex darwinianus (Forel)
Iridomymex itinerans (Lowne)

FORMICINAE
Camponotus claripes Mayr
Camponotus maculatus Forel
Camponotus nigriceps (Smith)
Camponotus testaceipes (Smith)
Stigmacros brooksi McAreavey

SPHECIDAE
Podagritus imbelle (Turner)
Podagritus leptospermi (Turner)
Tachysphex mackayensis Turner
Tachyspex paucispina Pulawski

COLLETIDAE
Euryglossa cupreochalybea Smith
Euryglossina hypochroma Cockerell
Euryglossina lynettae (Rayment)
Euryglossina perpusilla (Cockerell)
Euryglossina pseudoatomaria Exley
Hyleoides zonalis Smith
Hylaeus alcyoneus (Erichson)
Hylaeus amiculus (Smith)
Hylaeus violaceus (Smith)
Hyphesma atomicans (Cockerell)

HALICTIDAE
Homalictus bremerensis (Rayment)
Homalictus megastigmus (Cockerell)
Homalictus sphecodoides (Smith)
Homalictus urbanus (Smith)

APIDAE
**Apis mellifera* L.

Appendix 4
LICHENS RECORDED IN THE SOUTHERN FORESTS
(extracted from Sammy 1985)[a]

Aspicilia calcarea
Buellia subdisciformis
**Calcium sp.*
Caloplaca aurantiaca
Caloplaca ferruginea
Caloplaca fulgens
Catillaria sp.
Cladia aggregata
Buellia subalbula
Cladia ferdinandii
**Cladia schizopora*
Cladonia amaurocraea
Cladonia calycantha
**Cladonia cariosa*
Cladonia chlorophaea
Cladonia degenerans
**Cladonia furcata*
Cladonia kremelhuberi
Cladonia pityrea
Cladonia polydactyla
**Cladonia pyxidata*
Cladonia verticillata
Cladonia xanthoclada
Coccocarpia erythroxili
Collema glaucopthalmum
Dermatocarpon hepaticum
**Dermatocarpon lachneum*
Ephebe lanata
Fulgensia bracteata
Graphic scirpta
Heterodermia obscurata
**Hypogymnia subphysodes* var. *austrerodioides*
Hypogymnia subphysodes var. *subphysodes*
Lecanora atra
Lecidea decipiens
Lepraria candelaris
**Lepraria membranaceae*
Leptogium phyllocarpum
**Menegazzia caesiopruinosa*
Menegazzia platytrema
Ochrolechia subpallescens
Pannaria rubiginosa
Pannoparmelia wilsonii
**Parmelia cheelii*
Parmelia conspersa
Parmelia cooperi
**Parmelia erumpens*
Parmelia harrisii
**Parmelia imitatrix*
Parmelia perlata
**Parmelia pruinata*
Parmelia pulla
Parmelia reticulata
Parmelia rutidota
Parmelia subsumpta
Parmelia tinctina
Peltigera dolichorhiza
Peltigera spuria var. *erumpens*
Pertusaria leioplaca
Physcia endococcinea
Placopsis sp.
Porocyphus lichenelloides
Pseudocyphellaria aurata
Pseudocyphellaria australiensis
Pseudocyphellaria crocata
Ramalina celastri
**Rhizocarpon geographicum*
Sarcogyne clavus
Siphula coriacea
**Stereocaulon corticatulum*
Teloschistes chrysopthalmus
Thelotrema lepadinum
Thysanothecium hookeri
Thysanothecium hyalinum
Usnea barbata
Usnea scabrida
Usnea torulosa
Xanthoria ectanea
Xanthoria parietina

---

* Species recorded from Porongurup National Park but not from the rest of southern forests

[a]SAMMY, N. (1985). The distribution of lichens in Western Australia. *W.A. Herbarium Research Notes* No. **11**, 53-113.

Appendix 5

LARGER FUNGI RECORDED IN THE SOUTHERN FORESTS

(extracted from Hilton 1982 and 1988)[a,b]

CLASS ASCOMYCETES

Order Helotiales

* *Trichoglossum hirsutum*

Order Pezizales

Morchella elata
Peziza badia
Aleuria rhenana
Geopyxis carbonaria
* *Scutellinia scutellata*
* *Peziza whitei*
Plectania rhytidia

Order Sphaeriales

* *Daldinia concentrica*
Hypoxylon fuscum

Order Tuberales

Labyrinthomyces varius

CLASS BASIDIOMYCETES

Order Septobasidiales

Septobasidium sp.

Order Tremellales

Aporpium caryae
Tremella mesenterica
* *Tremelloscypha australiensis*

Order Agaricales

Agaricus placomyces
Agaricus silvaticus
Amanita ananaeceps
Amanita grisea
Amanita murina
Amanita ochrophylla
Amanita punctata
Amanita subalbida
Boletus multicolor
Boletus nigerrimus
Gyroporus caespitosus
Cortinarius subarcheri
Crepidotus nephrodes
Crepitotus variabilis
Dermocybe splendida
Descolea maculata
Hebeloma aminophilum
Tubaria rufofulva
Hygrophorus cantharellus
Hygrophorus laetus
Lepiota acutesquamosa
Lepiota cristata
Lepiota konradii
Pluteus pellitus
Lactarius clarkei
Lactarius eucalypti
Russula clelandii
Pholiota gummosa
Armillaria luteobubalina
Collybia abutyracea
Collybia butyracea
Collybia elegans
Conchomyces bursaeformis
* *Marasmius crinisequi*
Mycena austrororida
Omphalina chromacea
Omphalina ericetorum
Panellus ligulatus
Amanita dumosorum
Amanita griselloides
Amanita hiltonii
Amanita umbrinella
Amanita virgineoides
Amanita xanthocephala
Boletellus obscure-coccineus
Boletus sphaerocephalus
Cantharellus lilacinus
Coprinus disseminatus
Panaeolina foenisecii
Psathyrella candolleana
Cortinarius basirubescens
Cortinarius radicatus
Galerina autumnalis
Hygrocybe coccinea
Lepiota rhytipelta
Leucoagaricus naucinus
Melanophyllum echinatum

---

* Recorded from the Porongurup Range, but not from the southern forests.

[a]HILTON, R.N. (1982). A census of the larger fungi of Western Australia. *Journal of the Royal Society of W.A.* **65**(1), 1-15.
[b]HILTON, R.N. (1988). A census of the larger fungi of Western Australia Part II. *Journal of the Royal Society of W.A.* **70**(4), 111-118..

Paxillus panuoides
Paxillus muelleri
Naematoloma fasciculare
Pholiota highlandensis
Pholoita squarrosa
Anthracophyllum archeri
Lentinus dactyloides
Lentinus lepideus
* *Oudemansiella radicata*
Tricholoma eucalypticum

Order Aphyllophorales

Lentinellus hepatotrichus
Clavariadelphus juncea
Clavulinopsis helvola
Podoserpula pusio
Merulius corium
Ganoderma applanatum
Hydnum repandum
Phellinus scruposus
* *Chaetoporus euporus*
* *Piptoporus australiensis*
* *Piptoporus portentosus*
Polyporus mylittae
Polyporus pocula
Polyporus tumulosus
Polystictus brunneo-leucus
Pycnoporus coccineus
* *Trametes lilacino-gilva*
Lopharia crassa
Clavicorona pyxidata
Peniophora gigantea
Phanerochaete filamentosa
Laxitextum bicolor
Phellodon plicatus
Asterostroma persimile
Phellinus conchatus
Phellinus spadiceus
Phellinus wahlbergii
Dichostereum pallescens
* *Grifola colensoi*
Inonotus chondromyelus
Macrohyporia dictyopora
Melanopus infernalis
Phaeolus schweinitzii
Polyporus basilapiloides
Polyporus sclerotinius
Poria carneo-lutea
Stereum affine
Thelphora terrestris

Order Gautieriales

Austrogautieria manjimupana

Order Hymenogastrales

Hymenogaster albus
Cystangium sessile
Endoptychum agaricoides

Order Lycoperdales

Geastrum triplex
Castoreum cretaceum
Mesophellia glauca

Order Nidulariales

Nidula candida
Crucibulum laeve
Nidula emodensis

Order Phallales

Phallus rubicundus
* *Protubera canescens*

Order Sclerodermatales

Scleroderma flavidum
Scleroderma flavidum macrosporum
Scleroderma verrucosum

Order Tulostomatales

Calostoma fuscum

Appendix 6

REPRESENTATION OF PLANT ASSOCIATIONS IN NATIONAL PARKS AND RESERVES IN THE SOUTHERN FORESTS

Plant formation	Plant Association	National Parks/Reserves in which association is known to occur
HIGH OPEN FOREST	karri (*Eucalyptus diversicolor*) and karri-marri (*E. calophylla*)	Donelly River (NR) Smiths Brook (NR) Strickland (NR) Eastbrook Rd (NR) Mt. Shadford (NR) Thames (NR) D'Entrecasteaux (NP) Pemberton (NP) Shannon (NP) Walpole-Nornalup (NP) Mt Frankland (NP) Boorara (CP) Dombakup (CP) Murillup (CP) One Tree Bridge (CP) Jervik (CP) Jardee (CP) Chester (NR) Leeuwin- Naturaliste (NP) Dalgarrup (NR) Porongurup (NP)
	yellow tingle (*E. guilfoylei*)	Walpole-Nornalup (NP) Mt. Frankland (NP)
	+ red tingle (*E. jacksonii*)	Walpole-Nornalup (NP)
	+ Rates tingle (*E. brevistylis*)	Walpole-Nornalup (NP) Mt. Frankland (NP)
	jarrah (*E. marginata*) and jarrah-marri	Smiths Brook (NR) Dickson (NR) Strickland (NR) D'Entrecasteaux (NP) Shannon (NP) Pemberton (NP) Greenbushes (NR) Milyeannup (NR) Blackwood (CP) Bridgetown (CP) Ellis Creek (CP) Mullalyup (CP) St Johns Brook (CP)

NP = National Parks, NR = Nature Reserves, CP = Conservation Parks

Plant Associations and formations are from Smith (1972, 1973) unless otherwise indicated

- Associations recognized by Christensen and Annels

+ - Most of the species occurrence

SMITH, F.G. (1972). *Vegetation survey of Western Australia, vegetation map: Pemberton (and) Irwin Inlet*. Department of Agriculture, Perth. W.A.

SMITH, G.S. (1973). *Vegetation survey of Western Australia, Vegetation maps of Busselton and Augusta*. Department of Agriculture, Perth. W.A.

Plant formation	Plant Association	National Parks/Reserves in which association is known to occur
	yarri (*E. patens*) (not well represented in reserves)	Blackbutt Reserve (NR) Greenbushes (NR) Mullalyup (CP)
CLOSED FOREST	marri or jarrah-marri	Blackwood (CP) D'Entrecasteaux (NP) Frankland (NP)
OPEN FOREST	jarrah or jarrah-marri	Bokarup (NR) Unicup (NR) Bolbelup (NR) Pardelup (NR) Thames (NR) Whicher (NR) Paget (NR) Greenbushes (NR) Mileyannup (NR) Scott (NP) D'Entrecasteaux (NP) Mt. Lindesay (NP) Shannon (NP) Mt. Frankland (NP) Rapids (CP)
	marri	D'Entrecasteaux (NP) Boorara (CP) Dombakup (CP)
	yate (*E. cornuta*)	D'Entrecasteaux (NP) Walpole-Nornalup (NP) Randell Rd (NR) Wamballup (NR) Torbay (NP)
	bullich (*E. megacarpa*)	D'Entrecasteaux (NP)
	flooded gum (*E. rudis*)	Lake Muir (NR)
	W.A. peppermint (*Agonis flexuosa*)	D'Entrecasteaux (NP)
	Warren River cedar (*Agonis juniperina*)	Shannon (NP) D'Entrecasteaux (NP) Walpole-Nornalup (NP) Pemberton (NP) Mt. Frankland (NP) Torbay (NP)
	river banksia (*Banksia seminuda*)	Shannon (NP) D'Entrecasteaux (NP)
	jarrah and *Allocasuarina fraseriana*	Mt. Frankland (NP) Walpole-Nornalup (NP) Shannon (NP)

Plant formation	Plant Association	National Parks/Reserves in which association is known to occur
WOODLANDS	wandoo (*E. wandoo*)	Perup (NR) Mt. Lindesay (NP) Mordalup (NR) Unicup (NR) Randell Rd (NR)
	flat-topped yate (*E. occidentalis*)	Mt. Lindesay (NP)
	Melaleuca preissiana-Nuytsia floribunda	D'Entrecasteaux (NP) Shannon (NP) Mt. Frankland (NP)
	Agonis flexuosa	D'Entrecasteaux (NP) Walpole-Nornalup (NP) Leeuwin-Naturaliste (NP)
	Banksia littoralis	Walpole-Nornalup (NP) D'Entrecasteaux (NP) Shannon (NP) Frankland (NP) Mt. Lindesay (NP) Perup (NR)
	# + Red flowering gum (*E. ficifolia*)	Walpole-Nornalup (NP) Mt. Frankland (NP)
	jarrah-marri	Unicup (NR) Bolbelup (NR) Perup (NR) Mt. Lindesay (NP) Walpole-Nornalup (NP) Mt Frankland (NP) D'Entrecasteaux (NP) Shannon (NP) Whicher (NR) Mileyannup (NR)
	jarrah and*E. haematoxylon*	Whicher (NR)
	# bullich	D'Entrecasteaux (NP)
	# *Allocasuarina fraseriana*, jarrah and Albany blackbutt (*E. staeri)*	Mt. Frankland (NP) Mt. Lindesay (NP)
LOW CLOSED FOREST	jarrah-marri	D'Entrecasteaux (NP) Walpole-Nornalup (NP)
	{*Melaleuca rhaphiophylla* {*Melaleuca cuticularis* {*Melaleuca preissiana*	Shannon (NP) Lake Muir (NR) Mt. Frankland (NP) Mt. Lindesay (NP) D'Entrecasteaux (NP)
	# + *Melaleuca baxteri*	Torbay (NP)

Plant formation	Plant Association	National Parks/Reserves in which association is known to occur
	Agonis flexuosa	Leeuwin-Naturaliste (NP) D'Entrecasteaux (NP) Walpole-Nornalup (NP) Quarram (NR)
LOW OPEN FOREST	*Melaleuca rhaphiophylla* *Melaleuca cuticularis* *Melaleuca preissiana*	Shannon (NP) Lake Muir (NR) Mt. Frankland (NP) Mt. Lindesay (NP) D'Entrecasteaux (NP)
	Agonis flexuosa	Leeuwin-Naturaliste (NP) D'Entrecasteaux (NP) Walpole-Nornalup (NP) Quarram (NR)
	Banksia attenuata	D'Entrecasteaux (NP) Walpole-Nornalup (NP) Mt. Frankland (NP) Mt. Lindesay (NP)
	Banksia ilicifolia	Lake Muir (NR) Frankland (NR) Walpole-Nornalup (NP) D'Entrecasteaux (NP) Shannon (NP) Mileyannup (NR) Unicup (NR)
	Banksia littoralis	Walpole-Nornalup (NP) D'Entrecasteaux (NP) Perup (NR) Mt. Frankland (NP) Yarnup (NR) Lake Muir (NR)
	jarrah-marri (not well represented in reserves)	D'Entrecasteaux (NP) Walpole-Nornalup (NP)
	# *Banksia attenuata*, jarrah, Albany blackbutt and *Allocasuarina fraseriana*	Mt. Lindesay (NP) Mt. Frankland (NP)
LOW WOODLAND	*Melaleuca rhaphiophylla* *Melaleuca cuticularis* *Melaleuca preissiana*	Lake Muir (NR) Perup (NR) D'Entrecasteaux (NP) Mt. Lindesay (NP) Shannon (NP) Walpole-Nornalup (NP)
	Agonis flexuosa	D'Entrecasteaux (NP) Walpole-Nornalup (NP) Leeuwin-Naturaliste (NP)

Plant formation	Plant Association	National Parks/Reserves in which association is known to occur
	Banksia ilicifolia	Lake Muir (NR) Unicup (NR) D'Entrecasteaux (NP) Mt. Lindesay (NP) Shannon (NP) Walpole-Nornalup (NP) Mt. Frankland (NP) Whicher (NR)
	Banksia littoralis	Lake Muir (NR) D'Entrecasteaux (NP) Mt. Frankland (NP)
	jarrah-marri	Bolbelup (NR) Mt. Lindesay (NP) Mt. Frankland (NP)
	mountain marri *(Eucalyptus haematoxylon)*	Whicher (NR)
	# *Banksia attenuata*, jarrah and Albany blackbutt	Mt. Lindesay (NP)
	# *Eucalyptus decipiens*	Perup (NR)
LOW OPEN WOODLAND	*Melaleuca preissiana*	Lake Muir (NR) Unicup (NR) D'Entrecasteaux (NP) Shannon (NP) Mt. Frankland (NP)
	Agonis flexuosa	Leeuwin-Naturaliste (NP) D'Entrecasteaux (NP) Walpole-Nornalup (NP)
	Banksia ilicifolia - mixed spp.	Lake Muir (NR) Whicher (NR) D'Entrecasteaux (NP) Mt. Frankland (NP)
	Banksia littoralis	Shannon (NP) Mt. Frankland (NP)
	jarrah	D'Entrecasteaux (NP)
CLOSED SCRUB	*Melaleuca viminea*- *Melaleuca lateritia*	Lake Muir (NR) Perup (NR) Mt. Lindesay (NP) Unicup (NR)

Plant formation	Plant Association	National Parks/Reserves in which association is known to occur
	Melaleuca incana	Perup (NR) Lake Muir (NR)
	# *Melaleuca huegelii*	Leeuwin-Naturaliste (NP)
	# *Agonis juniperina*	Shannon (NP) Mt. Lindesay (NP) Mt. Frankland (NP) D'Entrecasteaux (NP) Walpole-Nornalup (NP)
	# *Kunzea ericifolia - Pultenaea reticulata* - on quartzitic sands. (not well represented in Reserves)	D'Entrecasteaux (NP)
	Dryandra sessilis	D'Entrecasteaux (NP) Leeuwin-Naturaliste (NP)
	# *Homalospermum firmum- Agonis linearifolia*	Shannon (NP) D'Entrecasteaux (NP) Mt. Lindesay (NP) Mt. Frankland (NP)
	# *Acacia littorea*	Leeuwin-Naturaliste (NP) D'Entrecasteaux (NP)
	# *Phlebalium anceps- Oxylobium lanceolatum*	D'Entrecasteaux (NP) Walpole-Nornalup (NP) Mt. Frankland (NP) Mt. Lindesay (NP)
	# *Agonis linearifolia*	Leeuwin-Naturaliste (NP) D'Entrecasteaux (NP) Shannon (NP)
	# *Agonis parviceps*	Shannon (NP) D'Entrecasteaux (NP) Walpole-Nornalup (NP) Mt. Frankland (NP) Mt. Lindesay (NP)
	A flexuosa	Leeuwin-Naturaliste (NP) D'Entrecasteaux (NP) Walpole-Nornalup (NP) Torbay (NP)
	# *E. calcicola-E. cornuta- E. magacarpa*	Leeuwin-Naturaliste (NP)
OPEN SCRUB	*Agonis flexuosa*	Leeuwin-Naturaliste (NP) D'Entrecasteaux (NP) Walpole-Nornalup (NP) Torbay (NP)

Plant formation	Plant Association	National Parks/Reserves in which association is known to occur
	Agonis flexuosa- *Dryandra sessilis*	Leeuwin-Naturaliste (NP) D'Entrecasteaux (NP)
	# *Allocasuarina humilis-* *Melaleuca viminea*	Perup (NR)
	Hakea varia- *Hakea prostrata*	Perup (NR) Unicup (NR) Lake Muir (NR)
	Melaleuca viminea	Perup (NR) Lake Muir (NR)
HIGH OPEN SCRUB	*Kingia australis* association (not well represented in Reserves)	Whicher (NR) Mt. Lindesay (NP)
CLOSED HEATH	# *Melaleuca thymoides,* *Lysinema ciliatum,* *Calytrix leschenaultii,* *Hibbertia pulchra* - on granitic sands.	Perup (NR) Mt. Frankland (NP) Mt. Lindesay (NP)
	Melaleuca viminea	Lake Muir (NR) Perup (NR)
OPEN HEATH	samphire (not well represented in Reserves)	Walpole Nornalup (NP) Lake Muir (NR) Quarram (NR)
	Jacksonia sp., *Acacia littorea* *Loxocarya sp.* *Olearia axillaris, Acacia sp.* *Scaevola spp.*	D'Entrecasteaux (NP) Leeuwin-Naturaliste (NP) Leeuwin-Naturaliste (NP) D'Entrecasteaux (NP) Walpole-Nornalup (NP) Torbay (NP)
	Myrtaceous spp.- *Astartea facicularis-* *Baeckia spp.-* *Petrophile spp.*	Walpole-Nornalup (NP) D'Entrecasteaux (NP)
	# *Gastrolobium villosum-* *G. spinosum-* *Dryandra armata-* *Leucopogon pulchellus*	Perup (NR)
LOW SCRUBLAND	*Olearia axillaris-* *Acacia spp.* *Scaevola spp.*	Leeuwin-Naturaliste (NP) D'Entrecasteaux (NP) Walpole-Nornalup (NP)
	Mixed sclerophyll scrub on ironstone	Perup (NR)

Plant formation	Plant Association	National Parks/Reserves in which association is known to occur
HERBLANDS/SEDGELANDS		
A number of different herbland and sedgeland associations exist. They are not detailed by either Smith or Beard and have yet to be described. However Smith recognizes three major types in the area.		
	Sedgelands on the south coast	D'Entrecasteaux (NP) Shannon (NP) Walpole-Nornalup (NP) Mt. Frankland (NP) Mt. Lindesay (NP)
	Sedgelands on the Margaret River	Not in Reserves
	Open herbland - marram tussock grassland (Introduced)	D'Entrecasteaux (NP)
SWAMP VEGETATION	# Cladium swamps (*Baumea* sp.)	Bolbelup (NR) Cowerup (NR) Lake Muir (NR) Unicup (NR) Perup (NR)
	Juncus pallidus swamps	Gingilup Swamp (NR) D'Entrecasteaux (NP) Shannon (NP) Walpole-Nornalup (NP) Mt. Frankland (NP) Mt. Lindesay (NP) Torbay (NP)

All of the remaining herblands and sedgelands which are within State forest have been recommended for inclusion in the Road, River and Stream Reserve system (see text).

APPENDIX 7
DESCRIPTION OF THE BOUNDARY OF THE SOUTHERN FORESTS

The southern forests as defined in this publication comprises the area within a line drawn from the west coast at latitude 33°39', to a point of longitude 116°05', north-east of Grimwade. From there a straight line runs approximately south-east to the Blackwood River. It then follows the boundary of CALM's Southern Forest Region (See Anon 1987), generally heading east to the point where it joins the Albany Highway and then through to the south coast at Albany. The boundary of the study area then follows the south-western coastline back to the startpoint at latitude 33°39' (see map in Fig. 1).
